ON THE WAY WITH GIS

Student Edition

TONI FISHER

loca+e
PRESS

Credits & Copyright

On the Way with GIS
Student Edition

by Toni Fisher

Published by Locate Press LLC

Direct permission requests to info@locatepress.com or mail:
Locate Press LLC, PO Box 671897, Chugiak, AK, USA, 99567-1897

Editor   Gary Sherman
Cover Art   Madelaine Dingwall
Interior Design   Based on Memoir-LATEXdocument class
Publisher Website   http://locatepress.com
Book Website   http://locatepress.com/otw

Contents

Foreword to Students

Do you know how to read a story? Maybe that is a strange question, but there are pockets of people who have grown up without written stories. Maybe you are one of them. TV and DVDs have taken the place of books and the days of being read to sleep with a fairy tale may have passed you by.

Fairy tales are examples of stories that have taken a bad rap. Right along with nursery rhymes. Okay, lots of fairy tales are pretty politically incorrect and dated. Girls should not be waiting for a prince to come and save them, and who wants to be a princess anyway? And it's not just the youngest boy who needs to be clever and hardworking; regardless of your place in the family, you had better be prepared for the future. But there are still reasons to pay attention to fairy tales and to nursery rhymes. They aren't useless. Learning from stories still has merit.

Sometimes old ways work. Not everything old is bad, not everything new is better. Nursery rhymes are filled with nonsense, yes, but studies have shown that the rhythm of the language, not the content of the rhymes, helps babies with learning to speak. Fairy tales are a source of history, but more, they are representations of how we can learn and of how we did learn for centuries.

Stories were told aloud, before they were written, mostly because not everyone could read. Stories allowed the listener to learn vicariously. Children, it would be hoped, did not have to fight a dragon to know that it was a dangerous proposition. They did not have to be caught by a wicked witch to know that, if someone told them the forest was dangerous, perhaps they should avoid the forest. Stories allow the teller to give information to the listener without there being an authority and an obedient audience. The story puts the action in a framework where it stays separate from the teller and the listener. It gives the listener an opportunity to experience without danger and without being admonished or being forced into behaviours. The consequences happen to the characters in the story. The listener can draw their own conclusions and decide how they would behave, given similar circumstances.

So, for a moment, forget that fairy tales are not stellar examples of teaching prowess, for content, and think about stories in general. People still like stories. TV and DVDs tell us stories that we don't have to read. We sit, passively, and take the story in. On the other hand, when we read, we have to fill in the gaps with our imaginations. We're more active in the story. A good story affects us emotionally. We get caught up in the story and live with and through the characters. Do we learn as much from stories we watch as we do from stories we read? Have a think about that. Discuss it in a group, argue both sides and see what you come up with.

Maybe some of the stories in films and TV teach us something, but most of the writers of film and TV scripts aren't really in the business of trying to inform or teach, they just want to entertain and provide a product that will be economically viable. There is still a market for people wanting to tell stories, using a modern medium, and to teach, at least for very young audiences. But what is being taught may not help us as we get older. After all, in the old days, the children had to grow up quickly. They needed life lessons at a young age. If you don't read stories, or access the right kinds of stories, where do you get the life lessons from? And do you know how to get them from a story if you haven't been exposed to them?

This book, On the Way with GIS, uses stories and it may approach learning from a way that is different from what you would expect. You might think that teaching and learning are all about having your teacher tell you exactly what to do and you doing what you are told. That does work, to a certain extent, but it is a short cut. You learn content, but you don't learn how to learn.

In fairy tales, the audience learns how to know who are the evil characters, who are the good characters, what the good characters have to do to win or succeed, what situations to avoid and how to get out trouble once they're in it. There are cues for the audience. To learn from the stories in this book, you will have to read them carefully to pick up on the cues. The stories are not meant to teach you particular content, although you will interact with them through content, they are meant to focus your learning on how to learn.

What are you supposed to do to succeed in the chapter? Where might the pitfalls be that will lead you astray? How will you master the interfaces that you have to approach? Will you be the character who does the basic minimum to get by, or will you be the hero? Can you engage with your learning in the same way that you would engage with a computer game? Can you want to win?

And now, for an admission. I am an avid reader, but I am not a risk taker. I want to know that the books I read are going to end well. I want to be able to relax as I read, even if the action becomes intense. Accordingly, I admit that I am a person who reads the end of the book first. I'm perfectly happy knowing the outcome, so that I can enjoy the journey.

Some people are horrified by this. They would never read the end of the book before they get there, but there are reasons for wanting to know the end before you get started. In learning, this is called "Stating the Learning Outcomes". The idea is that you should know what you are aiming for as you get started in your learning. This is usually what you will be tested on. It can help, but learning outcomes tend to be a bit dry and can be incomprehensible since you haven't done the learning yet. Too many people just skip over them.

In this book, let me suggest that as well as reading the learning outcomes below, you read the last chapter, not that you do the work in the last chapter, but that you read it, early on. You might want to think about the story in the last chapter as a way to approach the learning throughout the book. Maybe it will provide you with a happy ending. Maybe you will be a great learner! I won't tell you more. Read the story and see what it can teach you about how to learn.

Learning outcomes: By the end of each chapter, you will

1. Be able to complete the exercises to stated specifications using GIS software

2. Be able to evaluate and discuss issues surrounding the use of GIS software and data

3. Be able to evaluate and formulate learning strategies for effective learning

Now for more about being an effective learner and why it is important to be one. We used to learn skills to do jobs. Once learned, a person could do a job for a lifetime. Maybe it was not the most exciting way to spend most of your waking hours, but it was satisfying, for some, to have jobs that were predictable and secure, jobs where a person worked their hours and then went to their homes and another part of their lives.

These days jobs are rarely predictable, secure, or long lasting. Automation has replaced many kinds of work that required one repetitive type of skill. Jobs that are not being automated are the kind that require thinking and learning. People who can take disparate pieces of information and create something new with them are needed. People who can take a body of work and break it down through analysis cannot be replaced by automation. These are the kinds of abilities that jobs are more likely to require in the future.

How can you become the kind of person who is good at creating or good at analysing? That is your goal. Learning facts and keeping them in your memory is a good skill because it exercises your memory. That may be important in the development of your learning skills. But it won't be the facts that are so valuable, it will be the skill of a good memory. How can you use this book, how can you use GIS software, to develop skills that teach you how to learn, teach you how to be creative, teach you how to be a good analyst? Think about that as you read the stories and work through the chapters.

The questions at the end of each chapter are deliberately worded to encourage your thoughtful answer. Yes and no, very short, right or wrong, factual answers are not what is being sought. If you are looking for a question that has only one answer, you won't find it here. Instead, the questions ask what you think, backed by what you have read about. There is always controversy. People will always disagree about best methods and about best interpretations of outcomes. You will be asked to provide answers that weigh the controversies. You may not have a definitive opinion as to who is right, or more likely to be right, but you should be able to appreciate both sides of an argument. This will indicate that you have achieved some kind of understanding of the topic.

Do you want to learn how to learn? If you do, great, you are ahead already. If you haven't thought about it yet, here is your chance. If you read help files and watch online videos to learn, that is one way to get started. If you try first and then read and watch later, that is another way, but be aware that you may have to unlearn before you learn. That isn't a bad thing, necessarily, as you will always have to unlearn and relearn as you go through life. Things change quickly. As Alvin Toffler put it, "The illiterate of the 21st century will not be those who cannot read and write, but those who cannot learn, unlearn, and relearn."

Learning how to work with software is an important skill, learning how to work with GIS will be useful. Learning how to learn is priceless. Information is available easily, but knowing what to look for, how to look for it and what to do with it once you find it, that's what you need to know.

Ask yourself, is my learning style successful? Do I have an easy time learning? Can I learn quickly? Or, looking at it from another perspective, am I worried or afraid when I am told I am going to learn something new? Do I get frustrated when presented with something new that needs to be done? If the prospect of learning is not something you look forward to, make changes. If you do not have a strategy that allows you to approach learning without fear, take the time to develop one. If the way you approach learning does not have good results, and is not enjoyable as it takes place, change your approach. There are lots of different ways to learn. Find the one that suits you and make it better every time you learn. Learning is going to be part of your life, always. Learning how to learn successfully may be the most important thing you ever do.

Everyone starts from a different place. If you have an anxiety attack when something new is introduced, or if you have a temper problem when things don't go as you anticipated, you may want to improve how you deal with learning and your feelings. If you are easily bored or you find that your mind is wandering while you are in school, you may want to explore how to develop your own challenges and how to take advantage of every opportunity that presents. Regardless of where you start, where you get to and how you get there, that is what is more important.

You may want to record your journey. It may be steady or it may have periods of quick development and then plateaus. The point is to make the journey and to be on one. So, get started and get On the Way with GIS!

1. Resolutions about Resolution

Learning Tip One: For a complete perspective, read the whole chapter all the way through before you start. It's never a waste of time to be prepared.

"You're going to love this!" my friend ran up to me and grabbed my arm. I stopped suddenly on the busy sidewalk to listen and was rammed by someone from behind who was not watching where he was going, flicking through his mobile as he walked.

"Sorry," he muttered, not looking up.

"What am I going to love?" I asked, glaring after the person who had bumped me.

"Our next trip!"

Oh dear, I thought, here we go again. I had barely recovered from the last adventure my friend had organised. Not to mention the trip before that. My friend was an enthusiastic outdoors person. Not a natural athlete, but always willing to try. In fact, there was little my friend was not enthusiastic about. Sometimes it was a bit much to be with someone who was interested, no, fascinated, passionate, about everything. I, on the other hand, was a cautious person with very little athletic talent—okay, truth be known, no athletic talent. And I had learned to be even more cautious after the first time we went on a trip.

It was a bicycle trip. My friend had rented special road bikes with panniers. We had all the latest, lightest equipment. The trip was to be a 500 mile adventure along the coast. "We might see dolphins and whales," enthused my friend. "We'll stop in charming villages for tasty meals. We'll sleep under the stars."

It did sound good. I liked the ocean, watching the waves wash peacefully onto the beach. I loved the thought of seeing whales and dolphins. Good food was always appreciated, too. I found myself nodding in agreement as I listened. I could see myself parking my bike, sitting at a table, soaking up the sun while I ate delicious food that I didn't have to make for myself. Living in the city, I did not often get to see the stars. I could lay outside the tent and look up at the Milky Way every night before I drifted off to sleep with the sound of the ocean soothingly playing in the background. "How long do you think it will take us?" I asked.

"We can do it in nine days," answered my friend, "but I think we should plan for ten. Look, here is the route."

My friend pulled out a map and showed me the start and finish and the stages along the way. I was regaled with a list of where we would stay, including photos, and the places where we would refuel with food during the day. It did look good and it did look as if he had thought of everything. To complete the preparation, we trained for 2 weeks in a gym. We rode the exercise cycles for thirty miles every day. I read a lot of books while I rode my exercise bicycle.

The day came and we started riding. The bicycles were great. Our gear was great. After a few days, my legs were getting to be great, well, as great as they could be, given my lack of athletic tendencies. It was then that I discovered the part of the planning that he had forgotten. The fourth day was a roller coaster.

Yes, the route was 500 miles. Yes, he had the map that we would follow, but he had not made a profile map. Five-hundred miles we could have done. But we were not counting on the hills we faced that day, and once at the top, the rolling country I saw beyond. Hill after hill as far as the eye could see. A profile, that shows elevation of the route, would have shown us what we were really up against. It wasn't a squiqqly line on a map, it was a squiggly line in three dimensions. From here, to there and up and down, up and down. More up than down, it felt to me. At the end of the day, some of the hills had felt to my leaden legs as if they were Himalayan mountains. The exercise cycle had never prepared me for this. And I am still trying to forget the driving rain and howling winds we encountered a bit later on. A peaceful ocean? That had been only in my dreams. The ocean had been a wild thing, most days, spraying spume into the air and even onto the road at times, not to mention the smell of dead fish and rotting kelp.

Once again, I looked up at my friend's excited face. A new trip...hmmm, I thought. My muscles twinged and I unconsciously began to rub the bug bites that I still felt. The bugs had eaten me alive when I had tried to see the stars on the few nights when the rain had stopped pouring.

"This doesn't involve bicycles, does it?" I asked, suspiciously.

"No, no," said my friend. "Better than bicycles."

"Okay," I said, still suspicious. "What about skis?"

The second trip my friend had organised for us was a backcountry skiing holiday. "Four days in a winter wonderland. You'll love it," he said. "We'll be shushing down slopes in our backcountry skis, gliding over glistening white snow, untouched by anyone. And every night a mountain hut to sleep in, complete with feather duvets and hearty meals."

I liked the winter. I liked snow. I thought about the fun times I had had as a child playing in the snow. When we had enough snow, we would lay in it and make snow angels, snow men, snow forts even. Snow was fun. Cold, but fun. I reasoned that skiing on dry snow was better than riding a bike in the rain.

"What about the maps?" I had asked.

"Right here."

There were several maps on the table in the room where my friend had introduced the ski trip plans to me. I looked at the top one. The route and the huts were clearly marked. But I was not convinced so easily.

"What about a profile?" I asked. "We might shush down the slopes, but we'll have to do some climbing up as well."

"Right here," my friend said, obligingly shifting the maps.

The map underneath the first had shown the profile of the route. At least elevation had been taken into account.

"How many miles a day?" I asked. "How much elevation?"

"Well, I calculated the distance while adding in the elevation that would happen for each day. See," my friend pointed, "we will ski less distance on the days we have some major hills."

I could see myself in the mountains in the quiet of the winter. The snow muffling all sounds. Only the footprints of tiny animals marring the perfect white surface. I could see myself sitting by a cozy fire, the logs crackling, eating a hot meal and drinking hot chocolate at the end of a bright, winter day.

My friend and I trained for two weeks at the gym on the skiing machines. We threw in some rowing and some elliptical training too, just so that our legs and arms were extra strong. I felt very fit and was ready to go.

We had started skiing early one frosty morning. The sky was blue, the sun was bright. My skis flew along. At the end of the day, the hut came into view just as we were tiring. All was going as planned. However, at the end of the third day, just as dusk was falling and I was beginning to look anxiously ahead, anticipating our arrival at the mountain hut, my friend stopped suddenly in front of me on a gentle slope. I tried, but I couldn't stop. My friend saw me coming and fell into me so that we both went over backwards.

"What was that for?" I asked, spitting snow out of my mouth, brushing it from my eyelashes and digging it out from beneath the cuffs of my coat. Snow was beautiful, but it was too cold for rolling around in so late in the day. When I looked up from my snow removal efforts, I could see the lights, bleary through my wet lashes, of the hut down below us. I was looking forward to getting there.

"Ummm," my friend played for time. "There's a problem." My friend pulled out the map and the profile map.

"What kind of a problem?" I asked worriedly.

"Look ahead, just where we were going to ski," my friend pointed with his pole, sitting down in the deep snow. "See, this is where we are on the route, and this is the profile. See," my friend repeated, a bit puzzled, "the slope here is supposed to be quite gentle but it's more like a cliff. A crevasse, I think it's called. Looks about 40 metres wide." He edged forward, an inch at a time, "Wow, is it deep!" he whistled in appreciation of the dramatic cleft.

I looked at the sheer drop in front of me. I looked from the top to the bottom. I shivered at thinking what would have happened if we had gone over. Not even a mountain goat would try to undertake such a challenge. Then I looked longingly at the hut, on the other side. I could see the chimney smoking. I could practically smell the fresh bread that would go with my soup. Soup and bread that would be cold by the time we figured out a way to get to them.

"Well, we're not going to jump over it. Okay," I said. "We'll have to go around. It will take us a few more hours, probably," I said, trying to look for ways around, "but we can do this." Thank you, thank you, exercise equipment, I thought to myself. My friend's ideas were doing my health a world of good, but I wanted my stomach to be happy too. "Let's go!" I kicked off at a good pace.

When we arrived, we asked the keeper of the hut what had gone wrong with my friend's map and profile. "I think I see what the problem is. A profile map is made from a Digital Elevation Model," she explained. "Elevation points are often taken from air photographs or satellites. The spacing of the points is taken from the resolution of the image. A 30 metre resolution image means that the elevation is recorded at 30 metre intervals. If the resolution is 100 metres, or 500 metres, then only the changes between points at that distance are taken. The further the spacing, the less expensive the data is to produce. Do you know what the spacing on your points is?" she asked.

My friend wrinkled his nose, scrunched his forehead and scratched his head. "I remember something about that, when I was getting the profile produced for me," my friend admitted, "but I didn't pay much attention and I'm not sure I quite understand."

"We use 30 metre resolution around here, at a minimum," she said. "There are lots of steep hills, even cliffs, and people need to know what they are getting into in the back-country. Sometimes knowing can mean the difference between life and death."

I nodded and poked my friend, accusingly. "Do you have any better information?" I asked her.

"Of course, let me find it for you and let me explain more about metadata and resolution. Better yet, let me show you and show you how to work the data. I've got something here that will help."

She took us to her office and invited us to sit around her laptop. She opened QGIS and added a digital image. She showed us the metadata on the image and pointed out the size of the pixels, the resolution. She explained that if the resolution was say, 60 metres, the profile would look different.

"You make it look so easy," said my friend, "but I'm not so sure I can really understand until I work with it myself. I know QGIS a bit, so I should be able to work it out, if you'll let me have a go."

"That is an excellent idea," she agreed. "How about you?" she asked me. "Do you want to give it a try as well?"

"Actually," I paused, wrinkling my forehead, "I'm pretty tired. I don't know that my brain would take it in. I'll let my friend, here, do the learning for both of us. Okay with you?" I asked.

"Sure," said my friend. "I'll tell you all about it as I work it through."

"I know you will," I said, with mixed feelings.

"Okay," she said. "Let me get you started. First you need a DEM to work with. You can get one from this short tutorial: `http://loc8.cc/otw/gistutor_sample_raster`. It will give you a DEM to work with and make a profile, as well as setting you onto the next stage."

> The DEM and tutorial are also available in the data package, available at: `http://locatepress.com/otw_files/otw_data.zip`.

"What's the next stage?" asked my friend.

"Creating another DEM with a coarser resolution so that you can compare the two. You get started. Come and get me when you get stuck."

"First, I get the DEM," said my friend. "Then I make the profile of it. Next, I make the coarser DEM and then another profile. Sounds easy."

That's my friend, I thought, ever the optimist. I slouched in the corner in a very soft chair, enjoying the heat of the room and listening to my friend talking through the steps.

Follow along with the steps with the raster from the tutorial, entitled *How to sample raster datasets using points in Quantum GIS (QGIS),* or choose to download a different raster from the Internet. If you choose another dataset, you may want to clip just a small piece of it to work with so it will process quickly. Use the Vector->Geoprocessing Tools->Clip menu item. With this you can make a profile and then start on the next stage.

My friend read out loud from the Tutorial, "Copy and paste the URL for the dataset into a browser. `http://loc8.cc/otw/dem_data`. Save the sample raster elevation dataset. (Figure 1.1).

Figure 1.1:

It downloads to my download folder as a zip file. Click on the file, click on *Extract*. Save the file to a folder (Figure 1.2).

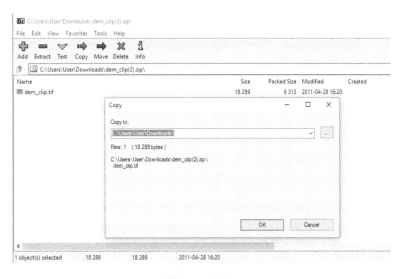

Figure 1.2:

In QGIS, click on *Layer->Add Layer->Add Raster Layer*. Now," mused my friend, "where was that profile tool?"

Just then the hut keeper stuck her head through the door. "All okay?" she asked. "Anyone want some tea?"

"Yes, please," I said.

"The profile tool?" said my friend who was single minded.

"Yes," she said, "I uninstalled it so that you would get to load it and see how it is done. Go to *Plugins->Manage and Install Plugins* and type `Profile` into the search box. There are several tools. Use the *Profile Tool*. Later, do try some others, if you want to experiment."

My friend worked away for a minute and then said, "Come look at this!" I sighed. Having a heart, my friend turned the laptop to me so I could see the screen without getting up (Figure 1.3).

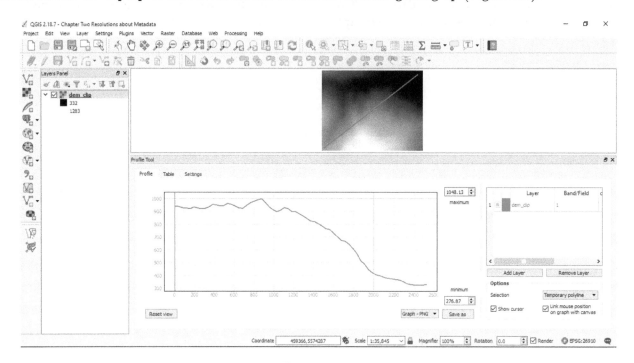

Figure 1.3:

"Look what I did! I clicked on *Add Layer* and then on *Temporary Line*. I drew a line from one corner to the other, double clicked to finish and there's the profile. This is amazing. With a digital image, I can now make profiles for us."

My friend clicked on *Save As* and saved the profile image, then closed the Profile Tool.

I could practically see my friend's brain whirling with new ideas and worried about what might be coming next. Would I be able to keep up? The gym is my friend, I murmured to myself, the gym is my friend, trying to instill confidence that I did not entirely feel.

"Okay, next is to see how the profile changes with a different resolution. The tutorial tells me how to find the resolution. Let me check that out."

My friend clicked on the layer for the image and saw the word, Metadata, in the Properties window that opened up. "Yes, 30 and 30, just as it says.

"Okay, now I want to test it out with 60, 60 for comparison." My friend followed the steps to generate Regular Points with 60 and 15 as the grid spacing, instead of 30 and 15 doing the following:

"Click the ellipsis next to the *Input Extent*, choose *Select Extent on Canvas*. Click on one corner of the image, drag and draw a marquee around the digital image. Fill in the numbers 60 and 15. Click on the ellipsis at the bottom to *Save to File*. Done!" See Figures 1.4 and 1.5.

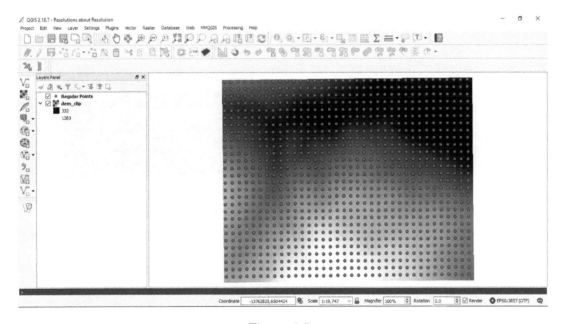

Figure 1.4:

Figure 1.5:

"Isn't this exciting!" my friend bubbled.

"Almost too much for me," I agreed with my eyes drooping.

"Click on *Plugins->Manage and Install Plugins*, type `Point Sampling` into the Search window. Click on *Install*," murmured my friend. "Find the little beast. Aha, it's under *Plugins->Analyses->Point Sampling*. Fill in the parameters..." (Figure 1.6, on the next page).

Figure 1.6:

My friend opened up the table to show me. "Fascinating. But I think we can find out more. Let me have a look..." (Figure 1.7).

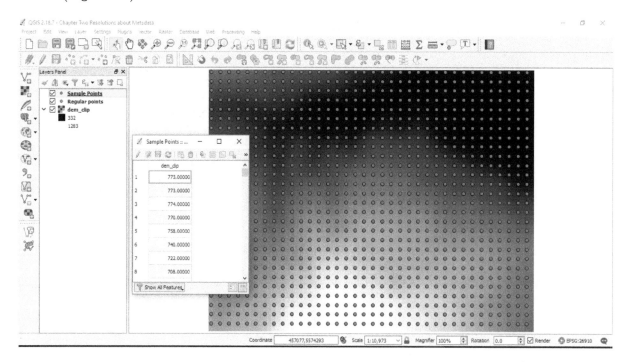

Figure 1.7:

"Yes, I thought so. *Vector->Analysis Tools->Basic Statistics for Numeric Fields.*" My friend was in a clicking frenzy. "Yes, look at all this data!"

"Wonderful," I murmured my encouragement.

"Now, how do I turn the Sample points into a new raster?"

Like magic, the hut keeper appeared. Tea for me and answers for my friend.

"How's it going?" she asked.

"Wonderful!" he said. My friend's 'wonderful' didn't sound quite the same as mine had. "But I need a hint. I want to turn the Sample Points into a new DEM with a coarser resolution. How do I do that?"

"Click on *Raster->Conversion->Vector to Raster*," she said, as though it was common knowledge.

My friend hesitated a moment. "Can you remind me why that would be?"

"Points, lines and polygons are vectors. A DEM is going to be a raster. You want to convert the points, the vector layer, to a raster. Does that make sense?"

"Thank you," said my friend.

The hut keeper put the tea on the table in front of me. I noticed, gratefully, that there were some chocolate biscuits beautifully arranged on a plate beside the tea. Statistics had excited my friend beyond belief, but the biscuits were things of beauty to me.

I nibbled. I sipped. I watched my friend clicking and inputting (Figure 1.8).

Figure 1.8:

"New raster! Check the *Properties->Metadata*. Yes, 60 and 60!"

I wondered if I should save any biscuits for my friend. No sense interrupting the flow, I rationalised as I continued.

"Complete!" shouted my friend. "Look at the two profiles!" (Figure 1.9).

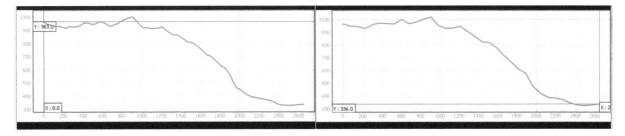

Figure 1.9:

"And you can see the difference. The one on the right is the 60 metre resolution. It is missing some of the detail that the 30 metre one has."

The hut keeper came in and looked at the two images. "Great job," she said.

"I've really learned a lot," said my friend. "I understand so much more about resolution and profiles. I resolve to pay much more attention to metadata."

"I've learned too," I said. "I resolve to check that you have checked the metadata."

"And I've learned as well," said the hut keeper. "Usually I show people how to do this. Half way through my demonstration I can see their eyes glazing over as they think about what they are going to do next or what's for supper." I knew just what she was talking about. She continued, "Letting you do it by yourself, without my help, has shown me that this is a much better way to teach and learn. Congratulations on your learning."

Thank goodness it was time for supper and then for bed. I was completely tired out. My friend was still talking about profiles until I fell asleep and then started right back up the next morning.

The hut keeper had saved our trip and we thanked her effusively as we left. We reconsidered our route and we did have fun but I was still wary about any great, new ideas that my friend might have for this vacation.

Suddenly I came back to the present and out of my memories. My friend had hit me on the shoulder. I started forward, merging with the surge of the crowd on the street. For a moment, I swore I could still see snow under my feet, and feel my aching legs. I had been really deep into my thoughts. My friend said, "Are you listening? How many times do I have to repeat myself?"

"Sorry, what?" I asked.

"I said, you're going to love the idea for this new trip."

Some people would have asked, "Where do you think we should go? What kind of trip is it?" but I was careful, as usual. I asked, "Do you have elevation data? How good is the resolution of the image? Is it aerial or satellite? Have you checked the metadata?"

"Forget the old ways," he waved his hand dismissively. "I have the best and the newest. I have LIDAR! We're going on a diving expedition!"

Oh, my goodness, I thought. I wonder how the gym can prepare me for that?

––––––––––––––––––––––

To complete this chapter, your goal is to:

1. Create two profiles, one with 30 metre resolution and one with a coarser resolution. Create a screen capture of them. Be prepared to demonstrate how you made them.

2. Follow the instructions for A-E and provide written answers for the questions in bold italics.

2a. Have a look at other data sources, such as the one in the link below.

Accessing and downloading data can be complicated and time consuming. Learning how to do it efficiently takes practice. The video link assists you with how to access the data and download. The Earth Explorer site holds the data. You do not need an account to explore the data, only to download it. If you like, you can use the data from the Earth Explorer website instead of the tutorial data suggested, but consider using the Clip function, as mentioned previously, for any other datasets so that you do not have a long processing time.

Watch the video at

`https://www.youtube.com/watch?v=0YPFegTcL4w`

Access the website at

`http://loc8.cc/otw/usgs_explorer`

Explore the data available and click on the icon to look at the metadata, being sure to click on the FDGC format button for the full description. The video will tell you to access the Data Sets Tab, Digital Elevation, SRTM 30 m. You can also check other kinds of DEM data sets.

Be aware that if you decide to download data, the datasets are quite large and may take time. If you want to work with the data, you may want to Clip a piece of the raster rather than work with the whole image.

Is there any metadata you have discovered that you think is particularly useful or interesting?

2b. Search for and read about Metadata and GIS. Use your own search results or consider these links:

`http://loc8.cc/otw/metadata`

`http://loc8.cc/otw/gis_wiki`

`https://www.fgdc.gov/metadata`

Which of the people in the story do you think would be good at metadata? Why? What kind of characteristics do you need to be a good producer of metadata? Would you be good at keeping metadata? Why or why not?

2c. Read about OGC and metadata standards.

`http://www.opengeospatial.org/`

`http://loc8.cc/otw/ordinance_standards`

`http://loc8.cc/otw/geospatial_standards`

Standards take a long time to develop. Standards make it easier for people to work together and ensure quality information, but not everyone is in favour of detailed standards, arguing that standards impede creativity and change.

What do you think about standards? Are they necessary, good to have, or a hindrance?

2d. *Why could you not use the 30 metre resolution image to create a 10 metre resolution and profile?*

2e. *What is LIDAR data? What is it used for?*

2. A Turn of the Tale

Learning Tip Two: When software doesn't work as you expect, it is usually operator error, much as you might want to hope that the software is to blame. Consider trying a different tactic to solve the problem. Experiment methodically with different approaches, watch a video of the software being used or read the help file. Don't just keep clicking in the same way, but, most important of all, don't give up!

Insufferable, that was what her cousin should be named. Not Isabella. How much did Ella have to take, listening to her all weekend, having to be polite, having to offer her the first and best of everything from soup to nuts, from the best bits of fluff to curl up in to the best spots in the sun under the trees while they listened to grandpa's stories? And why? Because she was a guest. Fine, but she was so full of herself she didn't need any more food and she could have floated on her own fluff, there was so much hot air inside her. She thought she was so good, being from the big city. She made sure Ella, her country cousin, knew that she was second class all the way.

If I hear someone say City Mouse and Country Mouse one more time, I'm going to stuff their tail down their throat, thought Ella as she sat eating supper with her extended family.

"Well, you know," said Berg, the huge mouse on her left who liked to be called Ice Berg, but whom everyone secretly called Ham Burg, "it's the old story about the City Mouse and the Country Mouse, isn't it, Ella?" He jabbed Ella in the ribs. Too bad Ham was just too big. His tail would have to stay where it was, although Ella cheerfully fantasised about what he would look like with it poking back through his teeth. She was lost in her thoughts for a moment.

"Ella, are you listening?" asked her mother. She repeated, "Gisella!" Her Mum only used her full name when she was really annoyed. It snapped Ella out of her reverie. "Will you please pass the rose hips to Bella?"

"Oh, thank you, Auntie Ruth," Bella simpered. "We eat so much jam in our house. We rarely see rose hips, or any primitive food anymore. It is so good to remember what food used to be like before we found different sources."

"I like hips," said Ella's Dad. "Fresh, crunchy, full of goodness. You can practically taste the sun."

"Me, too," Ella was quick to add. "Give me fresh food any day."

Bella smiled around at the table, the look on her smug little face belying her smile. Ella could see that she was just humouring them.

Ella's mother sought to fill the breach. "Bella, I heard from your mother that you had a great idea to share with the other mice in your house."

"I did, Auntie Ruth, how kind of you to mention it." Bella looked at her paws as though she was shy and demure but Ella knew that she was just itching to have her successes broadcast as far wide as was possible. She'd have hired a loudspeaker like the one in the school, if she could. She'd have wanted the whole country to know. "Mouse alert, Bella is the smartest. Did you know she is a City Mouse?" Ella could just imagine the announcements. Her tail twitched in frustration.

"Do tell us all about it, dear," insisted Ella's mother, although she certainly didn't have to ask twice. Bella launched into her story.

"Well, picture this," she began brightly with her voice pitched perfectly to garner all of the attention at the table, "our house in the City had become so popular that we were endangering ourselves. I mean, everyone, just everyone wanted to live there. It really had become that alluring."

Ella heard her Dad sigh. She pushed the plate of rose hips closer to him so he would have something to chew during the forced listening session. He winked at her and they shared a smile.

"It was getting to be that we were crowded, all in one space, despite our huge accommodation. Our rooms are just massive, of course," her eyes flitted around our cosy surroundings as though to measure them. And clearly, based on her smug smile, they did not measure up. "But I discovered something to help us out."

"Something to do with a bit of paper, wasn't it?" murmured Dad.

"Ted, let her finish," admonished Mum.

"Yes, it was paper," Bella said, "but not ordinary paper. It was gridded paper."

I was listening. What was gridded? Of course, she wouldn't need to ask as Bella was sure to go on with her explanation, ad nauseam, Ella thought.

"Gridded paper has equal sized squares already drawn on it. Each square can represent a distance. So, if you have twenty squares, it could represent 20 metres of space, or 100 metres if each square represents 5 metres."

Berg was looking puzzled. Maths was not his strong suit, but Ella understood the concept. Interesting.

"And then, with the gridded paper and with some pencil leads that I found. Pencils, the wooden things people use, are too big for my paws," she held up her slim paws, smiling, "but the leads from inside them work very well. I used the lead to draw our house with each of the rooms. Using the gridded paper, it was drawn to scale. That's what they call it when you represent something big on something small, like a paper. I was able to show all of the families who had joined us where they could best set up their individual apartments inside the house. We certainly didn't want to be living in a huge colony. After all, we're not rats." She shuddered.

Quite frankly, I know some nice rats, thought Ella, but she kept her mouth shut.

"You know, Ted," said Ella's mother, "maybe we could use some gridded paper. We could plan our accommodation here a little bit better. I know we're not as popular a location as you are, Bella," said my mother, "but we do have quite a few folk here."

"Oh, I'm sure not for long," Bella laughed. "Everyone wants to come to the city, after all."

"Do you have any of the paper to show us?"

"As a matter of fact, I do." Bella pulled a small piece out of her bag and smoothed it onto the table. "You likely have never seen any of this before. In the city we have so many things, wonderful things."

Ella peered over at the paper. She'd seen that before. In fact, the school room had loads of it, but the students didn't use it much. Ella had seen them using their computers more than the sheets of paper in the shelves. She had an idea that computers might be a replacement for paper. Ella thought she would need to check that out. Wouldn't it be wonderful if the Country Mouse had something to show the City Mouse. Bella's house was in a shopping area, but Ella's house was in a school. People learn new things in schools. Maybe I could learn something too, she thought. *I'll show her*, she decided.

Ella waited and watched whenever she could. It took a month. Sometimes she couldn't stay awake when the teachers were talking. She saw that sometimes the other students had the same trouble. But she persevered, paying careful attention to what the students did on their computers. This is what she saw.

First, the students read about working with *Vector Layers* in the *User Guide, Working with Vector Data: Editing* and the *Training Manual Module: Creating Vector Data.* Ella saw that the students who read had less trouble with their work. She thought about that.

Then Ella watched the students practice using the tools in the software program, QGIS. Their teacher told them that when they understood how to use the tools, they would be mapping their school. Ella saw that the students sometimes made mistakes and had to start over again. She saw that some of them got angry and gave up, but some of them got angry and used their anger to motivate themselves to succeed. She thought about that, too.

Some of the students were using a Toolbar that they installed into their program from the *View->Toolbars* menu. It was the *Advanced Digitising Toolbar*. Others loaded a plugin for CAD tools. Ella watched how they used the help files and the tools.

The students started drawing their school, the yard around it, the roads leading to it, their houses. They made their screens come alive with representations of what their world looked like to them. Some students focused on buildings and roads. Others added parks and then detailed the parks with areas for trees and gardens. The maps depended on what the students cared about, what they were trying to show. Ella thought about that. The world was just the world, she had always thought, but now she saw that it meant different things to different people. Her cousin thought about her world as buildings and shops, but Ella's world was all about fields and trees. It wasn't that there weren't areas of parkland and lots of trees in the city, it was just that Bella didn't bother with them, she saw them as less significant than shops. In the country, there were houses and barns, schools and community halls, but, for Ella, mapping was about where to find the best mushrooms, where the flowers came first along the stream bank.

Ella thought hard about her mapping. In the evenings, she went onto the computers herself and experimented. It always made her giggle when she used the mouse. She couldn't wait to show her family, and her city relations, what she could do.

Instructions for exactly what to do for this chapter follow after the horizontal line below, but, as an overview, consider that Ella started drawing a small scale map, as illustrated in 2.1, on the next page. Then she zoomed into a larger scale and drew what was important to her in her immediate environment, as illustrated in 2.2, on the following page.

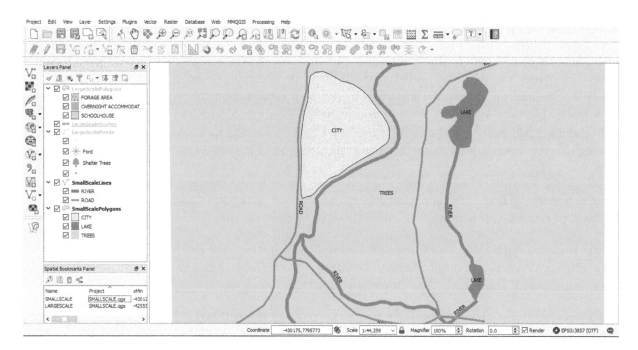

Figure 2.1: This map shows Ella's wider world in a small scale map.

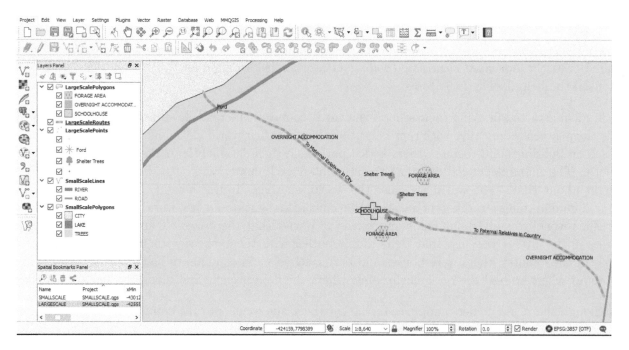

Figure 2.2: This map shows Ella's immediate environment in a large scale.

Small scale maps have scales that are larger numbers, such as 1:45000. Large scale maps have scales that are small numbers, such as 1:8000, as is illustrated. Note that the large scale layers, Large Scale Polygons, Large Scale Routes and Large Scale Points, are greyed out in the *Layers Panel* on the small scale map, Figure 2.1. Also note that the small scale features, the City, River and Road, Lake and Trees, frame the area of Ella's large scale map. Portraying the layers in this way is called *Scale Dependency*.

Your map will illustrate your use of scale dependency. Setting a bookmark for both scales, as is shown in the *Spatial Bookmarks Panel* at the bottom left of both Figure 2.2, on the facing page and Figure 2.1, on the preceding page, will make it easy for you to navigate between the scales. Be sure to set both your labels and your features to have scale dependency.

Do you remember how to add labels using a field in the Attribute Table? Do you remember how to add a field to the Attribute Table? If not, do a search in the Training Manual and the User Guide to find the appropriate help files.

Tasks:

1. You will make a map from your own perspective, illustrating what is important to you in your environment. To practice and improve your mapping skills, you will create a map using polygons, lines and points with layers for: `Small Scale Polygons`, `Small Scale Lines`, `Small Scale Points`, `Large Scale Polygons`, `Large Scale Lines`, and `Large Scale Points` (see Task 3 for details about Large and Small Scale). Practice snapping, labelling and selecting styles while creating shapefiles and displaying your layers. With the Advanced Digitising ToolBar (*View->Toolbars->/Advanced Digitising Toolbar*), you can reshape features to make them be exactly as you want them to be.

2. Make a list of the tools you use and the things you change away from the defaults. The more tools and the more default settings you change, the better your learning will be. Compile a report to show the tools and settings you have used.

- For 2a, aim for using at least 4 tools.
- For 2b, aim for changing at least 12 settings.
- Use screen captures to illustrate your report as shown in the examples. The undo and redo tool in the Advanced Digitizing Toolbar will assist with screen captures.
- Example for Tool List, Figures 2.3, on the following page and 2.4, on the next page.

2a. Tools:

2a.1 Node Tool, Before Move

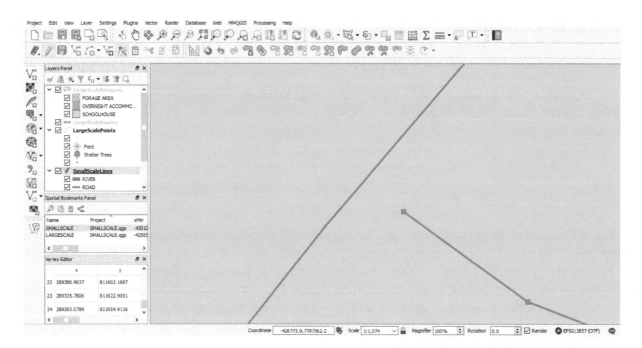

Figure 2.3:

After Node Tool Move, Figure 2.4

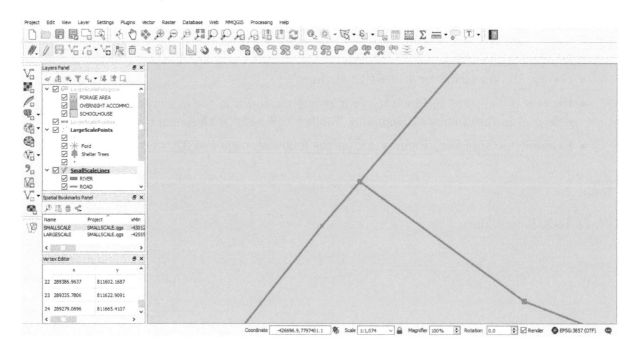

Figure 2.4:

(And at least three more tools to your report.)

Example of Default Setting Changes, Figures 2.5, on the next page and 2.6, on the facing page

2b. Default Settings Changes:

2b.1 Label Placement

Default is parallel

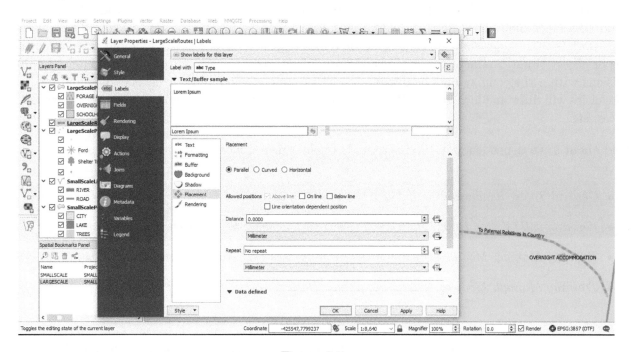

Figure 2.5:

Changed to Curved

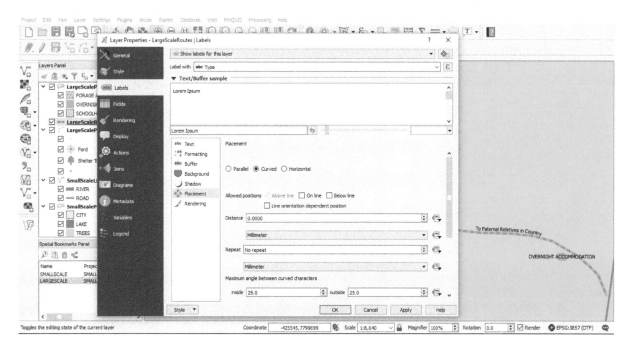

Figure 2.6:

(And add at least 11 more settings to your report.)

3. You will use scale dependency on your map.

Scale dependency means that certain mapped elements will only show up at a certain scale. Imagine a map of the world with all the roads on it and all of their names as labels, if they were all added. The map would be a black blob. There are too many roads. It would take forever to draw as well. The way online maps work is that things show up as they become meaningful, as a user starts to zoom in on a particular area or object. First continents may draw and have labels. As you zoom into a country, internal political divisions within the country may be drawn, as well as some of the larger cities. As you zoom farther in, more cities will draw and have labels. When you zoom into a city level, you will start to see transportation routes, etc.

Read about scale dependency in the User Guide (*Help->Help Contents*) in the following sections:

- *General Tools, Rendering*

- *Working with Vector Data, The Data Properties Dialogue, Scale dependent visibility*

- *Training Manual, Module: Creating a Basic Map, Lesson: Working with Vector Data*

- *Training Manual, Module: Classifying Vector Data, Lesson: The Label Tool*

At a large scale, such as Ella's immediate neighbourhood (Figure 2.2, on page 28), you will see details of the area in which she lives. Different mapped elements are needed for larger areas. The larger the area, the smaller the scale.

What you put in your map depends on your view of the world.

3. Over the Top or Over the Hill

Learning Tip Three: When there are lots of new terms to consider, write them down and write a definition beside them. You may never need to look at them again, but the act of writing will help you learn. Just taking the time, slowing down, gives your brain time to absorb ideas. Start slowly and then speed up. You'll end up being faster in the long run.

"Darling, you're here at last!" my Granny shouted at the top of her voice as she rushed down the front steps. I could hear her through the closed window of the cab and over the blaring radio that the driver had listened to all the way from the airport to my grandparents' house, thirty long miles away. It was definitely not a radio station I would have listened to but he seemed to enjoy it thoroughly. And now, as if my ears had not been assaulted enough, my Granny was about to do her best to complete the job, having announced my arrival to the entire neighbourhood. She was trailing yards of fluorescent purple and pink scarves as she bounded down the path to the road, her dayglo trainers flashing from beneath her swathe of scarves. She was more brilliantly attired than a dragon fly glinting in the sun.

There was no one like my Granny and I loved her dearly. A week at her house while my parents convalesced from their travel illness was going to be a rare treat. Who knew what would happen—certainly not me. My grandparent's house was bound to hold surprises.

Granny paid the cab driver and thanked him profusely. You would have thought that he had done her a tremendous favour out of the kindness of his heart, rather than just doing his job, but that was Granny, spreading cheer all around. I picked up my suitcase and raced after her as she bounded up the wooden steps to the front door that she had flung wide at my arrival.

"Pudge," she exclaimed, "She's here at last."

I heard a bit of a grunt and my grandad looked over the top of his paper. "Hello," he acknowledged and then winked at me.

"I still call him Pudge, in your honour," laughed Granny, recalling the error in my words I had made many years ago. Granny's fudge was famous to all who knew her, but when I was finally allowed a taste by my parents, who didn't believe in eating sugar, I got confused and called it pudge. Granny thought that was hilarious and began calling Grandad Pudge because of the vast amounts he could put back, despite his remaining as thin as a pole.

Grandad rolled his eyes a bit. "After you've settled in, do be sure to stop by for a visit," he said before retreating behind his paper once more.

"Come, come, dear, let me settle you in and fill you in on all the latest," Granny enthused as she started up the stairs. "I understand you're a bit busy right now with your studies. Your parents told me that I might be required to give assistance."

"Thanks, Granny," I said in my soft voice, not that she would notice. "But I don't think you'll be able to help. I have a project to do for school that is pretty technical." It wasn't that I doubted that my Granny was smart, because she was. She liked to tell everyone that she was a sharp as a tack, quick as a whip, no moss on her, the idioms went on and on, but I really thought that my new computer studies would be beyond her.

"Well, we'll see about that. Something to do with computers, isn't it?" she asked. "Wait until you see my new set up." She was now calling back to me from the upstairs hallway. "What's taking you so long, come along dear, we only have a week. So much to do."

Granny flung open the door to the guest room and stuffed me inside. "You know the drill," she said. "Unpack. Stuff your clothes, neatly, of course, if your parents ask, into the dresser. Clean up a bit if you feel the need and then come downstairs to the kitchen. I'll call your Mum and Dad and let them know you have arrived."

I did as I was told and made my way back to the kitchen where wonderful smells were emanating from beneath the swinging door. Granny was stirring several pots that held some kind of promise of lunch, but there was a big plate of fudge on the table.

"Do have some fudge, darling. I know you shouldn't before lunch, but this is a special day, isn't it? Your first day here. And then take some along to your Grandad. You know he can't resist."

I took the plate and went to see Grandad. The rooms he inhabited had a completely different feel from the ones that Granny splashed about in. Grandad was an oasis of calm. We shared a few pieces of fudge and discussed world affairs, or rather, Grandad told me about world affairs while I listened politely. Judiciously, I saved some room for lunch as I knew the portions would be generous and that Granny would have it on the table on the dot at twelve o'clock.

Granny insisted on a brisk walk following lunch. She and I raced through the streets. Despite being a bit out of breath, I managed to tell her about my QGIS project.

"I have to do something with raster data and analysis," I said by way of introduction, getting ready to explain to her what a raster was.

"Aerial photo?" she asked.

"Yes," I replied, somewhat surprised. But then, why would I be surprised. Granny was no ordinary Grandma.

"Orthorectified, georeferenced?" she continued.

"Ah, I'm not sure," I answered.

"Are you downloading an image?" she asked.

"I can," I explained. "I can also use something that I have already, my teacher said. I get to choose."

"Good," she commented. "I like teachers that give choices. I don't care for the ones that tell you exactly what to do, do you, and then get fussy when you think for yourself? How does that prepare you for life, that's what I want to know?

"Raster analysis is interesting," she continued. "There is a lot that can be done. You may not know this, but a long time ago I had a job where I worked with aerial photography. Now that was in the days before there were satellites and before there were computer programs that can instantly orthorectify an image."

"What is orthorectification?" I said, stumbling a bit over the word.

"When an image is taken by a camera from a plane, or a balloon, which, I might add, was even before my day," she chortled, "or a satellite, the part of the image that is perpendicular to the camera, the nadir, has no distortion. Anywhere else, the objects in the image will appear to lean away from where they are. Ortho means straight, or correct. So orthorectifying the image means to straighten it so that the scale of the area covered by the photo is all the same, as though it is all exactly below the plane, or satellite. In my job, I had data that was collected, on the ground, of elevation points and exact location. I took the aerial photos and stretched them, to fit the data on that was collected from the ground, moved them until I could make a photo mosaic of large areas. I had a pretty good eye. And, I was speedy."

I could imagine that was correct. I was panting a bit, trying to keep up with Granny's trainers as they flashed along the foot path.

"Now, if your image is already orthorectified and georeferenced, which means that you know exactly how it fits onto the world, you can get right onto doing analysis. Any ideas of what you might like to do?"

"Not really. I was going to do some reading first."

"Excellent idea," she said. "Read about what can be done. You have to take things one step at a time when you are learning. Make haste slowly, as I always say. Read, think, then let's talk. So many choices. I can't wait to help you out."

Granny is enthusiastic about everything, including learning, so she didn't want to take any opportunity away from me. She put me in the room where she and Grandad had their afternoon tea and their computers. Granny had a purple laptop she didn't want to share, so Grandad had an older PC by the big front window. Granny explained that her settings had taken her a long time to configure and that she was very particular about them.

I was going to use Grandad's PC so he showed me a few things about it. I was surprised to see that he had loaded it up with extra RAM. "It might look a bit old," he whispered to me, "but under the hood, it's all new." He winked. "Your Granny thinks I don't know about computers. But I'll tell you a few secrets later that will knock your socks off. Anyway, the download speed is fast, so anything you need to set up, go right ahead. Just in case you need it, here is my password." He wrote it down on a piece of paper—*OTH*. "Destroy it when you leave—I know I can trust you," he said conspiratorially.

"OTH?" I asked.

"Over the Hill," he laughed. "That's me."

I thanked Grandad and got right to work installing QGIS and downloading the manuals and tutorials I would need. Then I started reading and exploring, searching the Internet for data and ideas. By tea time I was ready with a plan.

"Granny, I was thinking about getting an image from the Internet, already georeferenced and orthorectified and then doing some Map Algebra on it."

Granny nodded, then said, "Ah, Map Algebra, that was something new for me. I didn't do that when I was working with aerial imagery, but I've read about it. Tell me more about what you are thinking."

"Well, I know that Map Algebra is just using the values in maps with operators like greater than, less than, equal to, things like that. So, I would be able to classify an image based on the values of the pixels and work with multiple rasters at the same time. I've seen some work done where soil type, precipitation, and slope of the land were used to predict the likelihood of a landslide occurring. Or, I thought about classifying orthoimages for land use."

"Supervised or unsupervised classification?" she asked.

"Both, maybe—what do you think? I haven't quite got it clear in my head about the differences."

"Both are good, but for different reasons and purposes. Let's say you have an orthoimage and you want to know what is in it. You can look at the numbers associated with the pixels in the image and group them. If the numbers are in the same group, they might be the same thing, maybe all the same kind of tree, or the same kind of plant, or an urban area. That is unsupervised classification. You can use unsupervised if not sure what you might find."

"On the other hand, supervised classification means that you train the classifier. You pick an area that you know is a tree species and specify that all pixels like it are going to be that kind of tree. Both supervised and unsupervised will result in a map that shows land use, or land cover. You can change the map from being a bunch of numbers of spectral reflectance, to one of being a class, a group, with a name."

"Spectral reflectance?" I asked.

"An interesting concept, that. Okay, so the sun is shining on a leaf. The leaf looks green to your eyes. That is because the wavelength of light that is green is not absorbed by the leaf, it is reflected by it. The leaf absorbs red and blue, not green."

"So, you mean the leaf is really not green? It is everything but green?"

"Yes, if you think that the colour of a leaf is what is absorbed then the leaf isn't green. We see the world's colour in a way that works predominantly with reds, greens and blues. Not all creatures see the world that way, of course. To insects, birds, other mammals, the world looks quite different. When we take images of the world for use in photogrammetry, we can use black and white film or colour film, or infrared. Some colour films have three bands, RGB, red, green and blue and some have more. You use RGB on computer screens."

"Spectral reflectance, supervised and unsupervised classification, map algebra. That's a lot of things that can be done with rasters," I said, trying to get my head around all the new words and ideas.

"It is," Granny agreed, "and it's so exciting! It all starts with an image that is orthorectified and georeferenced."

"Granny," I mused, "perhaps I should really just start with that."

"You mean orthorectification and georeferencing?"

"Yes. Then I could see what else I could do if there is time."

"Good idea," she agreed. "Start at the beginning and work your way up. Then you'll really understand what you are doing. Why not start with a scanned image and georeference it. I have just the thing."

Georeferencing is described in the *User Guide, Bookmark Plugins, Georeferencer Plugin* and in the *Training Manual, Module: Forestry Application, Lesson: Georeferencing a map*. The Training Manual's lesson is the one you will follow here.

In the sentence: "The sample data used in this module is part of the training manual data set and can be downloaded here", click on the hyperlink (downloaded here). Download the zip file and extract the forestry folder into your exercise_data folder.

http://loc8.cc/otw/qgis_forestry

As you are working through the lesson, pay careful attention to the crosshairs and the X,Y rationale so that you fill in the values correctly.

Granny explained, "This exercise will show you how to take an image and place it in the world. After you have placed it, you can change its transparency and show that you have put it in the correct location."

I worked away at it, georeferenced the image, and then showed them to Granny (Figure 3.1).

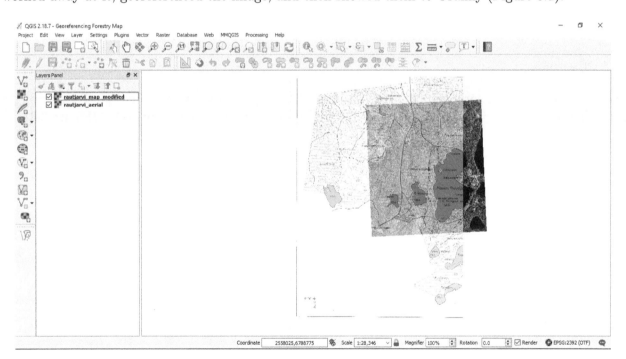

Figure 3.1:

"Good work," she said.

"But look at this Granny," I said. "If I add the original map into my project it doesn't show up." (Figure 3.2, on the next page).

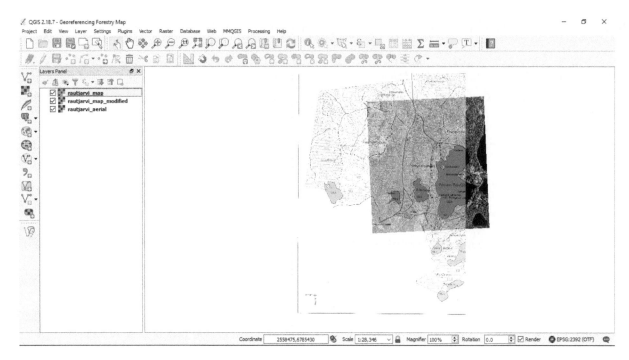

Figure 3.2:

"That's right. What you need to do is right click on the original, then click on *Zoom to Layer*."

I did what she said.

"See," said Granny. "The original image is somewhere far away from where it should be. Trying using the *Zoom Full* tool. Now you can't even see the images because you are zoomed out so far (Figure 3.3)."

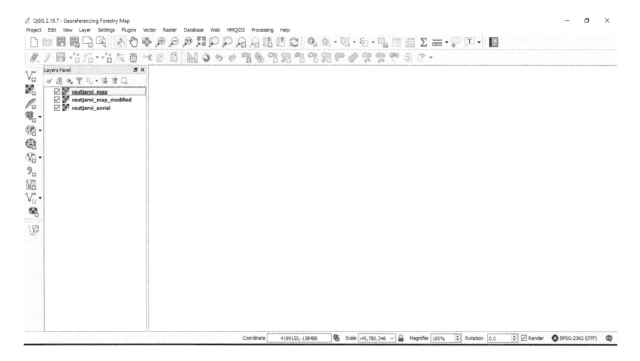

Figure 3.3:

"Go from one to the other, *Zoom to Layer* and *Zoom Full*, and look at the difference in the scale. Or you can use the *Zoom Last* and *Zoom Full* tools. But, you have the modified image in the right place now that you have georeferenced it. What do you think, will your teacher approve?" she asked.

"Oh, I think so. Anyway, I've learned so much that whether my teacher likes what I've done or not doesn't matter so much."

"I like how you think. You've discovered that learning is more important than grades. Well done, great job. But there's something else I want you to think about."

"What's that?" I asked.

"Georeferencing is pretty easy but knowing how to click the buttons in the program can sometimes be dangerous."

"Dangerous?" I thought Granny was kidding.

"Some people are really good with technology," she explained, "and some people are subject-matter experts. The trouble is that not very many people are good with technology and have the subject matter expertise. Because of that, we, I mean, our countries, don't always do the right thing."

I must have looked puzzled, because she continued, "It's kind of like knowing how to install a computer game, how to troubleshoot when the software breaks down, how to make the hardware and the software work together really well. That's the tech part. Then there's the person who knows how to play the game. The person who knows the rules. But their experience is hemmed in, is determined by the person who dictates how the software is going to work with the hardware." She frowned, I could see she was wondering if I would understand.

"I think I know what you mean," I said, substituting the ideas in her analogy. "It would be better if people who knew about good uses for Land Use maps also knew how to make them. If the subject-matter experts in Land Use were GIS experts too."

"Exactly," agreed Granny, "You know, as you've been working with georeferencing, you must have noticed how many projections there are to use. Just as images have distortions that need to be fixed, projections have distortions as well. Projections make geographers SADD."

"Sad?" I asked.

"S.A.D.D.," she said. "It's an acronym for Shape, Area, Distance and Direction. Those are the four elements that get distorted when we take the spherical world and try to put in on a flat map. In the Mercator projection, countries that are huge, appear to be small and vice versa depending on where they are in relation to the equator. Mercator distorts area. Sometimes that may have been used for political reasons, to promote propaganda and influence people's opinions.

"Web maps are the maps that most people use these days. The Mercator projection is the one most frequently used on the web. The choice of that projection was made by a technical person, not by a person who knew a lot about cartography. That means that the software companies have decided how maps are going to be portrayed and consumed. Geographers and scientists aren't entirely happy or satisfied by that, but there isn't anything they can do." She paused for a minute. "I don't know. Maybe if the two groups of people worked together on solutions, the subject-matter experts in cartography and the technologists, it would be better than working separately."

I nodded.

"Now, let me show you my newest game," she said, giving me a hug and casting aside the sombre mood. "Wait until you see my score. I wiped everyone out last night."

Granny went online and looked at the leader board.

"I can't believe it. Someone beat my score. She pointed at her alias, "OTT"."

"OTT?" I asked.

"Over the Top. It's what I call myself."

"Looks like OTH beat you and by quite a margin."

"OTH. Bet its short for Over the Hill. I'll show them. Humph!" Granny rolled up her sleeves.

I looked back into the room where Grandad was smiling while he read his paper. Over the Hill? Nah. Couldn't be.

After you have completed georeferencing the rautjarvi map, do these tasks and answer these questions.

1. Georeferencing is a first step in starting to work with raster images. The next step is to classify them so that analysis can be done. The Semi-Automatic Classification plugin is tremendously powerful although it requires some skill to use correctly.

1a. Install the Semi-Automatic Classification plugin. Open it from the SCP Menu Item.

1b. Read the Semi-Automatic Classification plugin manual, Chapter Three, Brief Introduction to Remote Sensing. Access the User Manual from the SCP Menu item once you have installed the plugin.

1c. In the QGIS GUI, have a look at some of the plugin tools that are installed with the plugin by clicking on them. Digital Image Processing requires considerable skill and is a fascinating field.

1d. In your own words, describe part of the plugin and what it is used for. Look for information about the tool you have selected in the User Manual.

2. Georeferencing depends on a projection. Read this article about how the choice of a projection chosen to enhance online consumption by technologists has affected cartography.

Available online at: `http://loc8.cc/otw/battersby`

Also available online at:

`http://loc8.cc/otw/web_mercator`

3. Think of an instance where technology has changed our world and where subject-matter experts are not being listened to. Describe it.

3a. What do you think would be better: if all subject-matter experts were technical experts or if technical experts and subject-matter experts worked closely together without one or the other dictating methodology or goals? Or, thinking of this in another way, do you integrate technology in your subjects at school, or do you have separate lessons for technology? What do you think works best? Describe what a class in your school would look like if it integrated technology and subject matter.

4. Play the game, The True Size found at: `http://thetruesize.com/`. Find at least one country that surprises you. What was it and why?

5. What happens when decisions about how we see the world are in the hands of commercial businesses, such as the companies that run the browsers or search engines that you use? How are you affected?

4. When Heat is not Enough

Learning Tip Four: Some tasks are easy to perform using tools, but understanding the science behind them takes more time. If something is easy to do, how much learning has really taken place? If it is easy, take the time to do extra and challenge yourself. Make the best use of your time. Fill it up with quality. Just getting it done is not the same as understanding what you have done.

"I'm worried."

"You're always worried."

"Yes, but this time it is about the children."

"It's always about the children!"

"I know, Ralph, dear. But listen anyway, please," Maggie begged.

"Alright. What is it about the children this time?"

"They're going missing."

"Missing? How can you be sure? I mean there are so many of them."

"A mother knows. I've been watching and counting and I've been worrying."

"There you go again. Worrying," I said. "Haven't we given them the best of everything? Haven't we given them a nice, quiet place in the country to start their lives?"

"That's just it," said Maggie. "They start here, but then they're gone."

"What do you expect?" I said. "They can't all stay here. How many are there, there must be more than a hundred."

"I know, but something is wrong. I know that some leave home, I know that some," Maggie stopped for a moment to deal with her thoughts, and continued, choking a bit, "are caught by spiders, or eaten by birds, or killed by humans, but lately something else is going on. There are too many who are simply missing."

"You mean, you've been counting?"

"More or less," she admitted. "And I've been checking all the usual spots. If I see their little corpses wrapped up in the webs..." Maggie broke off, trying to stifle her feelings. "If they are wrapped up, I know they're gone. If they are crushed on the window sills of the house, I know they've been swatted, but this is different. I know a few will be eaten by birds and I just won't find any evidence. They're gone. But there are just too many. It isn't the usual pattern."

"Maggie, we give them all we can," I tried to console her, to reason with her. "We give them a life in the country, a lovely, heated manure pile. What else can we do?"

"We need to teach them where to go, what to do, how to flourish. And I want to start teaching them where NOT to go."

"And how are we going to do that?"

"I want to make a heat map. I want to find out where the dangers are and I want to warn them before they fly out."

"Isn't the manure pile warm enough? What do you need heat for?"

"Not heat as in hot, Ralph," explained Maggie. "Heat as in hot spots of activity. In our case, hot spots of areas where there is an unknown danger. If we map the spots, we might be able to identify the danger and then prevent them from going missing. We could save their little lives."

"And how do we do that?"

"Let me show you." Maggie flew down to the ground where there was an open spot in the grass, covered in sand. "Pretend this is our field," she indicated the whole sandy expanse. "Now pretend that this is the fence and the long grass where the spiders tend to live. And this is the manure pile, the nursery for our dear little maggots. Over here is the house where the humans live. And here is the new water feature that the humans dug for the horses. If we map it all," she raced around making borders in the sand, "it looks like this."

Help Maggie out. You draw the map in QGIS. Here is an example of what your map might look like: Figure 4.1.

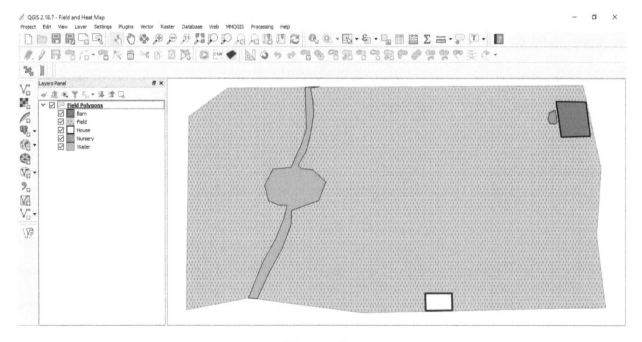

Figure 4.1:

"Now, what we want to do is to map all the locations where I have seen the young ones before...," she choked and could not go on.

"Before they went missing?" I asked.

"Yes, flying so beautifully one minute, gone the next," she lamented.

"It could just be birds, you know."

"I know the birds are dangerous, Ralph, but I think there is more to this than just birds. We haven't had a plague of swallows recently, have we?" she asked sarcastically.

"No," I agreed and then seeing how distraught she was, I said, "Come on, Maggie, show me what we can do. Let's see if we can figure this out."

"Okay," she nodded. "Thank you. Now, right now it doesn't matter where I put the locations, because I am only showing you how this works."

"Okay," I said. "I'll help you." We darted around in the sand for a bit, making six little dots each time we landed. When we finished, the sand looked something like this.

To make the map, digitize the points in QGIS. Click on Layer->Create Layer->New Shapefile to create a new shapefile for the points. Click on the Pencil tool to start editing. Click on the Add Features tool to add points.

You want a lot of points but you don't need to add them one by one. First digitize twelve points to represent the indentation for each leg of Ralph and Maggie as they land in the sand. Now use the Select by Polygon tool. Draw a polygon around the points, right click to end. Click on Edit->Copy. Click on Edit->Paste. The twelve points have become 24 but they are on top of each other. Click on the Move Selected Features tool. Click on the points, hold and drag to another location. Do combinations of copy and paste until you have enough dots to make a map with some crowded areas and some less crowded areas. It could look something like this (Figure 4.2).

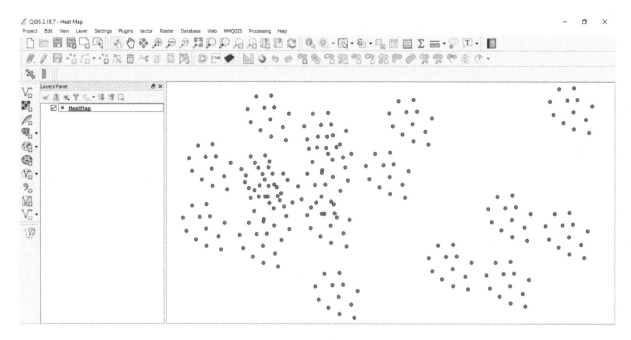

Figure 4.2:

"Now, if we use a heat map," continued Maggie. "A map that shows hot spots, places where there are a lot of events occurring, we'll be able to see if there is a pattern to the disappearances."

"You mean like here," I landed in a spot where we had added lots of prints in the sand.

"Yes," she agreed. "And here as well," she landed on another spot. "If we can map the places, in our field, maybe we can figure out what is happening. Maybe we'll see a pattern. Maybe we'll be able to warn the little dears as they fly out."

Use the dots you have now to practice making a heat map. Click on Raster->Heat Map. Use the points as the input file and save your raster with a name. Right click on the raster layer and choose Properties. Choose Style->Render Type->Single Band Pseudocolour. Click Classify. Click okay. You should have a heat map that looks something like Figure 4.3, on the facing page.

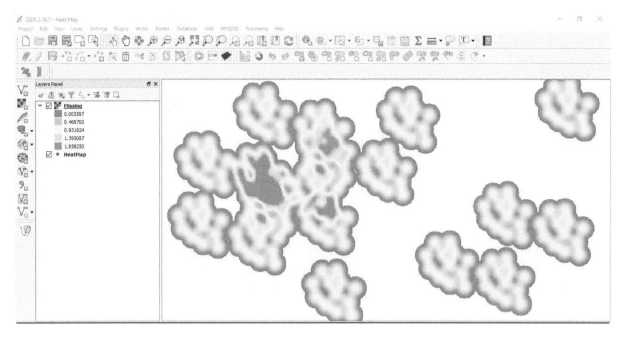

Figure 4.3:

A lot of maybe's, I thought. But, Maggie was right. We had a great location, we had a great nursery. We gave the little maggots all we could so that they would hatch, strong and healthy, but it was still our responsibility to see that their first flights away from home were safe ones. If Maggie was right, something had changed and we had to find out what it was.

"I'll get as many flies to help as I can. I'll send them out right away. You stay here. Collect and plot their information on the map when they come back. We'll see what we find."

"That's great, Ralph. But I think we need some help. I have someone in mind, in fact, but it means we have to go to the house."

Oh dear, I thought. What have I gotten myself into? The house? Past the swatters, past the fly strips... what was I supposed to do, who was I supposed to look for?

"Okay," I said. "I'll do it for you. I'll do it for the children."

I flew off. A little bit later, I landed on you.

"Please listen," I began. Thankfully, you weren't ticklish and you were open to an unexpected dialogue.

Now you will make the real heat map with the real data, as collected by the flies.

1. You have already drawn the field, with the features.

2. You will now add the points. The flies are bringing back locations and you are plotting them for Maggie and Ralph as quickly as you can. As you add the points of disappearance, and there are a lot of them, some spread throughout the field, likely from the voracious swallow attacks, you start

to notice a pattern. There are a lot of disappearances around the new water feature. There are especially a large number around the edges of the pond, near the water lilies.

3. To clearly illustrate the area of most danger, you will create a heat map, as Maggie and Ralph did with their practice data. You finish your map and get it ready to show to Maggie, Ralph and all the others.

4. What do you think is causing the disappearance of the young flies?

5. Read a bit more about heat maps, using the following suggestions, or any ideas of your own. Take notes and answer the questions.

Suggested Sources:

- `http://loc8.cc/otw/heatmap_tutorial`
- *User Guide, Bookmark Plugins/Heatmap Plugin*
- `http://loc8.cc/otw/heatmaps`
- `http://loc8.cc/otw/heat_vs_spot_map`
- Search for heat maps in a browser and select Images.
- Search for QGIS heat maps in YouTube.
- Search for information using phrases such as: Who uses heat maps?, What are heat maps used for?, Examples of Heat Maps, Misuse of heat maps, until you have a good understanding of the purpose and potential for heat maps.

Questions

1. Give three examples of heat maps. Who are they used by? How do they work?

2. Who is John Snow? What did he do with maps?

3. Heat maps are easy to create, using software. Can heat maps ever be problematic? Is creating a heat map always a good idea? What should you know before you use the software functions so that your heat map represents data correctly and is not misused?

5. Lost and Found

Learning Tip Number Five: Things don't always work out the way they should, or the way you expect the first time around. Is that a reason to stop? Should you just wait until someone rescues you? Maybe sometimes, but not always. Be intrepid. Learn fearlessly by making the extra effort to find a solution.

I was lost. I was tired and I was hungry. I was cold, too. My day had not improved from its dreadful start. I was planning on visiting a friend in a nearby city. Last night I put my train tickets in my pocket. I had packed a lunch and put it by the door. I set my trusty alarm clock.

The alarm clock no longer had my trust. It didn't ring. I woke up an hour late. My dog ate my lunch sometime in the night. My tickets were useless now that I was late. I would never be able to get new tickets at this busy time of the year. Everyone was travelling and the trains would be beyond full.

I ran to the station, leaned against the heavy doors, and was met with a seething crowd. I was bumped and jostled as I made my way through the station, often moving backwards just as much as I moved forward trying to make my way over to the Information Desk. By the time I arrived at the desk, I was nearly panting with the effort of moving against the flow of the crowd. I took off my coat and wiped the sweat from my forehead.

Did I mention I was cold? Why is that, you ask? I'm cold because I lost my coat when I was standing in line at the Information Desk to see if I could get a refund for my tickets.

"Of course not," the person at the ticket window informed me cheerfully, taking the time to tell me the kind of ticket I could have bought that would have allowed me to travel anytime, rather than the cheaper version I had purchased, which had now expired. "Next," she called to the person standing impatiently behind me.

Add that to the day, I sighed—an expensive trip that I wasn't on and a call I would have to make to my friend, explaining my absence. Not to mention the fun I would miss doing the things we had planned. Then I bent to pick up my coat from the floor where it had fallen from my arm when I was showing my tickets to the person at the desk. That was when I discovered the coat was gone, which is why I am cold. Useless tickets and a lost coat. Could this day get worse?

"Did you see my coat?" I asked the person behind me.

"No," he answered.

The person behind him shook her head as well. "The Lost and Found is over there," she said, pointing to an area over near the door where I had entered the station. To get there I would have to negotiate the crowds yet again.

I moved out of the line and studied the crowd and the layout of the hall. It looked to me that, if I angled over to the edge of the hall and then scooted along the far side of the pillars, I would be able to get there more quickly and with fewer bruises. I put my head down and pushed my way determinedly against the people, well-padded in their winter coats. At least the exercise would warm me up again. I really did want that coat back—it was one of my favourites.

A few minutes later I leaned against the far side of a tall, stone, pillar supporting the high dome of the station ceiling. There were not as many people on this side of the pillar as it was not part of the main flow for traffic. On the other side of the pillar, only a metre away, the train travellers were rushing by like a river in flood. As I looked at the crowd, I spotted a little boy clutching his father's hand. He looked up at me for a moment, with big eyes, and waved his teddy bear at me. I waved back just as he turned and dropped the bear. Before he could get his father's attention, they had been enveloped by the crowd and were pushed along with the flow. I reached out my hand to grab the teddy, but it was kicked by someone else, and it skidded across the floor just out of my reach. I jumped into the crowd, bent down, grabbed the teddy and jumped back. Hey, I was getting good at this. I had infiltrated the crowd, escaped it and evaded being elbowed or pushed.

On my way to get my coat from the Lost and Found, it would not be a hardship to carry the teddy bear and drop it off. Hopefully the little boy would be back to retrieve it. I looked at its well-worn fur and the tattered ribbon around its neck. This was a well-loved bear.

I could just make out a bit of a sign in the far corner of the hall: "nd." That must be it, I thought, "Lost and Found." I headed towards it, the teddy under my arm. Before I got there, I saw a hall on my right, narrow, dark, but at the end of the hall I could see a sign, quite low to the floor. "Lost and Found" was painted in small, gold letters on a white board. They should have put the sign up higher and made the lettering bigger, I muttered, as I started down the hall. It is almost as if they didn't want me to find the Lost and Found, I chuckled to myself. Then I thought, maybe they really didn't want anyone to find it. Maybe they didn't want the work of having to deal with customers.

Feeling annoyed, and still cold, hungry and tired, I reached the low, narrow door and pushed it open. It creaked as though it didn't admit many visitors. I peered inside and was met with a most astonishing sight. The room was filled with items, strewn in piles along the walls and in a row up the middle. It was like a colourful, lumpy quilt. Red, blue, purple, turquoise, green and gold melded into the patchwork.

A very short man peeked around one of the piles. "Can I help you?" he asked.

"Is this the lost and found?" I asked.

"One of them," he answered. "Which one are you looking for?"

"I didn't know there was more than one," I answered. "I lost my coat and I found this teddy bear."

"Then you've come to the right one," he answered. "There is a Lost and Found for luggage, and items with names printed or taped on. Then there is a Lost and Found for items that have no identification. When the Lost and Found can't return an item, because it has no identification, they throw it out. We retrieve it and we have a go at finding the right home to send it to."

"How do you find the owners if there is no identification?" I asked. "Or do you just hold it until a person comes?"

"You weren't listening," he reprimanded me sternly. "We take our job very seriously. I said, we send it to the right home and I meant send it. There are ways, you know."

"Tell me," I smiled, holding up the bear. "How can you find out where this little one belongs?"

"Do you speak Toy?" he asked.

I laughed. "No."

"Do you speak Cloth Object, perhaps Clothing, Blanket?" He narrowed his eyes and looked me up and down, appraising me. "Do you have any aptitude for footwear, perhaps?" he inquired.

I started to laugh again, but then my laughter froze. Several other extremely small people came out from behind other piles of lost belongings and looked at me, very seriously.

"Non-believer," muttered one.

"Give him a chance," said another.

"Let me tell you something," explained the man. "Items absorb things. When people are talking, they often mention their addresses. For example, I'll bet that the child who owned this bear has been told their address many times. Toys are easy. Absorbent material, parents who want to teach their child where they live. All it takes is being able to listen. Here, give me the bear."

I handed it over.

"220 Robin Street, Linburgh, UK. Easy."

I was fascinated. He handed the bear to a colleague. "Put the address in the geocoder and get this bear home," he said.

"What is a geocoder?" I asked.

"Geocoding is easy," he explained. "All you do is put the address in the software geocoding engine and then it creates a route to the bear's home. We'll get our delivery team to return him to his owner." He continued, "We work independently of the other Lost and Found. We like to give a little more care so we deliver, especially special items like this."

"Why not just give the address to the delivery team?" I asked. "Can't the delivery team read," I joked, looking around at the room.

No one smiled in return. Eyebrows were raised.

"We don't talk about the delivery team to just anyone," he said, frowning at me. "You of all people shouldn't be laughing. You don't even speak Toy. Anyway, their abilities are not something I'm going to share, right now," he answered. "On the other hand, if you can learn to speak Clothing, maybe I'll give it some consideration." He eyed me up and down. "Something tells me you have some talent. If you didn't, I don't think you would have been able to find our door. I have a feeling for things like that."

He turned away, bent over and reached into a pile. "Your coat," he said, handing it to me. "It is very happy that you came by. Come again some time. We could use the help."

I walked home. On the way I passed store windows. I stared for a moment at the clothes in one window. "No way," I told myself. I looked at the display in next window. "And I don't do shoes!" Still, it was interesting to think that I could help. But what could I do? They did geocoding, but I could learn something else and maybe show them. I decided to give it some thought.

So, what is geocoding? If you have an address, can you find a location? Of course, you can. It isn't magic. There are tools and applications on the internet that will allow you to do so. QGIS can do it as well and here is a chance for you to input the data in the way that search engines online do it and to do even more with it than the online programs allow you to do.

Read about geocoding and network analysis in the *QGIS Training Manual*. Bookmark *Vector Analysis, Lesson 7, Network Analysis*.

You need to load another plugin, the *MMQGIS* plugin, using the *Install and Manage Plugins* menu. It will appear as a new menu item. Read about the multiple functionalities of MMQGIS at `http://loc8.cc/otw/mmqgis`.

To use this Plugin for geocoding, you need to create a CSV file. CSV stands for comma separated value. Use a spreadsheet and create a file with your own address (Figure 5.1).

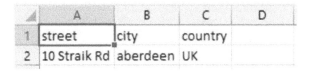

Figure 5.1:

Save the file as: CSV format. You may get a warning as you save your file, that formatting may be lost, but save it anyway, as it is safe to do so.

Open QGIS. Add a street map from the *Web* Menu Item.

Now run the geocoding engine. Click on *MMQGIS->Geocode->Geocode CSV*. Browse for your file. Notice that *Address Field*, *City Field*, and *Country Field* have mapped to your column names. Browse for places to output your shapefile and browse for a place to store the *Not Found Output List*.

Click *OK*. A new shapefile will be added to the *Layers Panel*. Right click on it. Select *Zoom to Layer*. You may have to use the Zoom tools to be able to see the point in context (Figure 5.2, on the facing page).

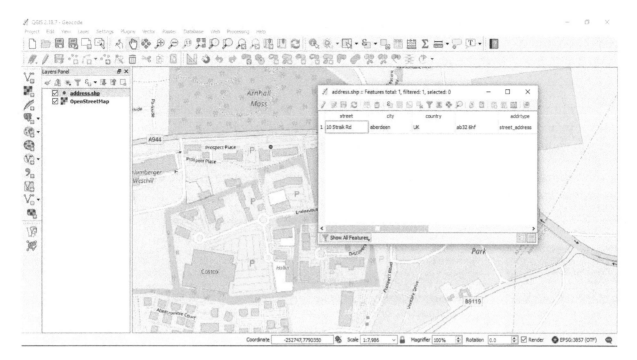

Figure 5.2:

Imagine you had 100 addresses in the CSV file—or 10,000. You can run the geocoder one time and it will place all the points on the map. Easy to do.

But what about getting from one place to another. What about calculating a route. There is a plugin that can do that too. Maybe it would help the delivery team. Maybe you could learn how and explain it to them. This is how it works.

Click on *Plugins->Manage and Install Plugins*. Scroll for the Roadgraph plugin (alphabetical order) and select it. Click on *Install*.

The plugin needs a road network. You could create a road network, by tracing a map, but that can take quite a bit of time to make a good one. Road networks are line shapefiles with attributes. Names of the streets, if the street is one-way or two-way and the allowable speed are just some of the information that could be collected and attached to the street lines.

On the other hand, a lot of GIS data is available for free on the internet, so you don't need to create your own. You may want to use the Ordnance Survey data which is free for download. Or you can try a different source for any other area in the world. You are looking roads in shapefile format.

On the Ordnance Survey page, `http://loc8.cc/otw/ordinance_downloads`, scroll to the *OS Open Map-Local* product, *Data type: Vector*. Use the map of Great Britain and select just one of the squares. Click the check box for *Download*. Scroll to the bottom of the web page and click *Continue*. Fill in the form so that you can receive the data via email. Note, your teacher may do this for you and provide you with a file if you cannot download.

The data will come as a zip file. Save it. Open the zip file. There are a lot of files. You only want the road files for now. Scroll down the list until you find the NO_Road. Shapefiles have several different pieces—they all must be in one directory on your computer for the shapefile to work. Make sure you select: .shp, .prj, .shx, and .dbf. Click on *Extract*. Save them to a directory.

Click on *Layer->Add layer->Add Vector Layer*. Browse to your `.shp` file. It will display in the Layers Panel.

Next, you will need to set up snapping. The routing tool allows you to click on a road as the start point of your route and click on another point as the destination. It can be difficult to click exactly on the road, even if you zoom in very close. Snapping makes it easier.

Click on *Settings->Options->Digitizing*. Fill in the snapping parameters for *Default snap mode*, *Default snapping tolerance*, and *Search radius for vertex edits* (Figure 5.3).

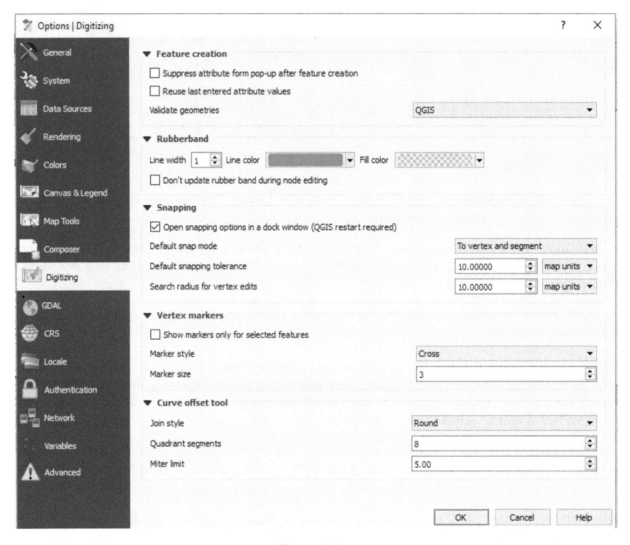

Figure 5.3:

Click on *Settings->Snapping Options*. A new panel will be added to your GUI. Fill in the *Snapping mode*, *Snap to*, and *Tolerance* (Figure 5.4, on the facing page).

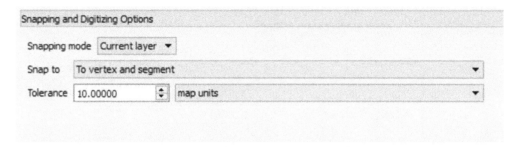

Figure 5.4:

Click on *View->Panels->Shortest Path*. Another new panel will show up. You can drag and dock this panel below your Layers Panel. Click on the *Title bar* and drag it down until you see the bottom of the Layers Panel. Let the title bar go and it will dock in place.

NOTE: If the Next Step, as detailed below, does not work, and you get a message that the plugin is not configured, even after you have configured the settings, read and follow these instructions before you try `Vector->Roadgraph->Settings` again!!

When you use someone else's data, you may not get exactly what you want. The data may have to be modified. If the data you download does not work in the plugin for creating a route, you may have to copy and paste the data into a new shapefile to prepare the data for the plugin.

Use the `Select by Polygon` tool. Draw a polygon around your roads, right click to finish. `Click Edit->Copy Features`. Click `Layer->Create layer->New Shapefile Layer`. Name your new file and save it. Click on the new layer in the `Layers Panel`. Click the `Pencil` tool to start editing. Click on `Edit->Paste Features`. Stop editing by clicking on the `Pencil` tool. Save. Now you have a shapefile that will work with the routing tool.

Next Step

Click on *Vector->Roadgraph->Settings*. Fill out the settings in the *Transportation* tab to configure how the shortest path will work. Fill in the *Topology Tolerance*. Your road layer will already be selected. If it is not, follow the steps in the **NOTE**, above. *Speed* should be *Always use default*. Click on the *Default Settings* tab. Fill in the *Speed* (Figures 5.5, on the next page and 5.6, on the following page).

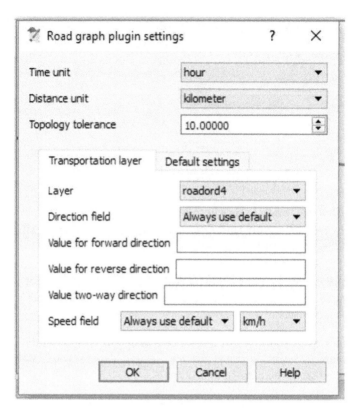

Figure 5.5:

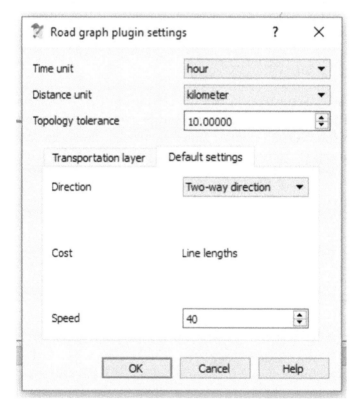

Figure 5.6:

Zoom into a place on your map. Have the scale be quite large (1:200, for instance). Click on the *Start* icon in the *Shortest Path* panel. Click on the road (Figure 5.7).

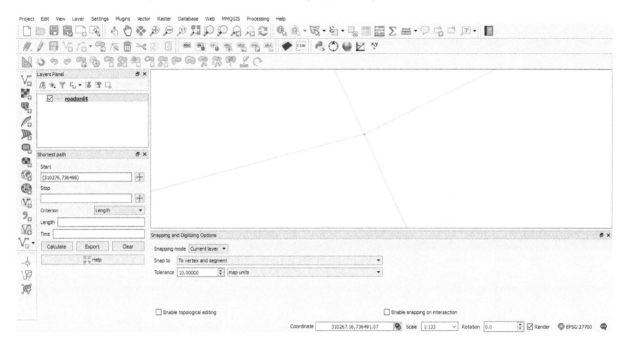

Figure 5.7:

Zoom out. As you work, consider *Zoom Last*, which is a handy tool. You can click it until you get back to where you want to be. Find where you want your route to end. Click on the *Stop* icon on the *Shortest Path* panel. Click on the road (Figure 5.8).

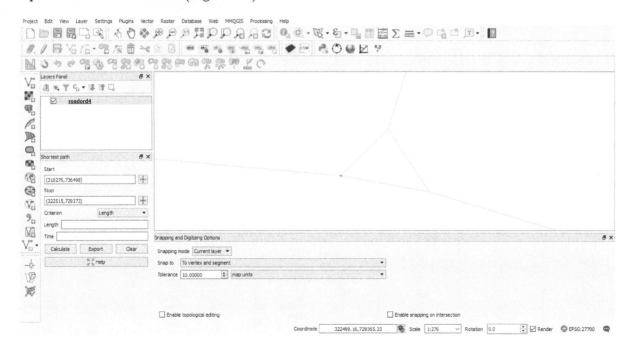

Figure 5.8:

Click on *Calculate* (Figure 5.9, on the next page).

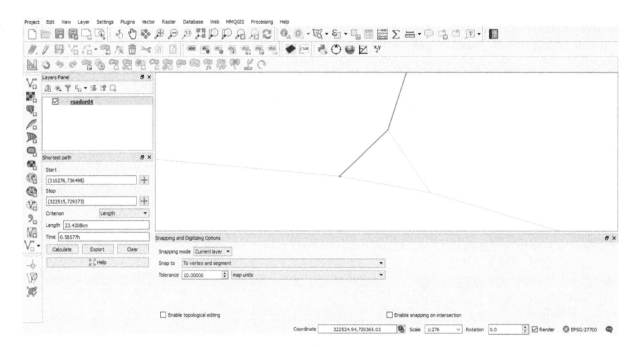

Figure 5.9:

Zoom out to see the whole route (Figure 5.10).

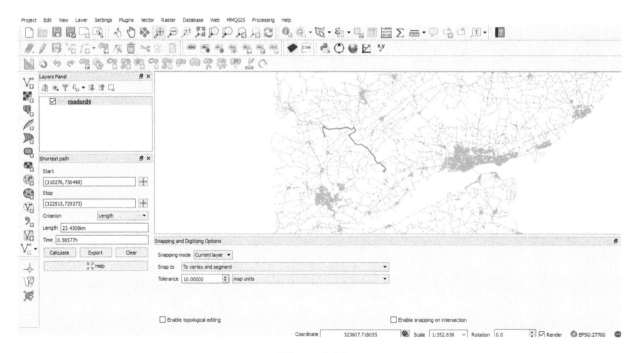

Figure 5.10:

Easy! Now let's go show the team at the Lost and Found.

The room was buzzing with activity when I pushed through the door the next day. "Back already?" said the person I had met the day before. "Did you already learn how to speak Toy?"

"No," I answered. "And I don't speak Cloth Object or Footwear either, but I do have something to show you."

He walked over, between the piles, carefully replacing the straggling items that had fallen to the floor. "Okay, what have you got?" he asked.

I whipped out my laptop and showed him how routing worked.

"Very nice," he said. "Good demonstration. Really useful. I can see the merit in this." He rubbed his chin for a moment. "It would be useful if our delivery team was using the roads. But our team has... errr... unique characteristics, shall we say." He looked up at me in a shy way. "Do you think you could do flight paths, by any chance?"

I knew I was going to have to have another think. Flight paths? What kind of delivery team did he have?

––––––––––––––––––––

Maps used to be drawn by hand and were works of art that took considerable time to produce and were not updated very frequently as a result. Now all commercial maps are created using a GIS where the updates can occur much more easily and the maps are often consumed online. Creating and printing a static map, that is often outdated as soon as it is printed, is only one option.

GIS underpins numerous commercial enterprises including delivery companies as well as government information repositories for transportation use and planning. Advertising companies exploit road networks to provide location based services. Our world uses GIS extensively with both good and problematic results.

1. Search for and read about how satnav maps are made.

    http://loc8.cc/otw/satnav_maps

2. Search for some potential problems with satnav. Describe one.

3a. What is a location based service?

3b. Give three examples.

3c. Can you think of any reason that location based services might be problematic?

4. You have read about the multiple functionalities of MMQGIS at http://loc8.cc/otw/mmqgis. Pick a function of MMQGIS that you think is interesting and that you might want to investigate further. Explain why you picked that function.

6. Turning Data into Information

Learning Tip Six: It's easy to make data look good when you use good software, but that doesn't mean the message conveyed by the data is accurate or valuable. Using data and software correctly is an ethical responsibility.

"Being reliable is important, isn't it?" asked our teacher.

Most people in the class nodded. I did.

"And when you want an answer, you want an objective answer, not an answer that is biased. You want the answer to be fair, not based on someone's personal feelings, would you say that is true?"

More nods and from me as well. Where was the teacher leading us, I wondered?

"Answers that are usually thought to be objective and reliable come from quantitative research. Quantitative research looks for answers, usually using numbers and statistical methods, on carefully controlled variables to test a hypothesis or theory."

We waited. Surely there was more to come.

"For quantitative research to take place, the researcher has to think of a question that needs to be tested. That might come from an idea, or an observation of an event. Then the researcher needs to have data to test if their idea is correct. The data has to be collected, found, or bought. It takes time and/or it takes money to get good data. But that is just the start. After the researcher has the data, it needs to be turned into information using statistical techniques that are reliable and reproducible." The teacher looked around the room at us. "Do you think you could do that?"

Some people nodded. I waited. I thought there would be more.

The teacher asked one of the nodders, "What part of the process do you think would be the hardest to do?"

"Some people might have problems with the statistics," Sophie answered, "but that's not that hard. There are lots of programs that can calculate the answers needed. I know a little about probability and I know that you can just click and get the right answers."

"That's true," agreed the teacher. "Anyone can click on a button and run a process. Do you think that is all you need to know?"

Sophie wrinkled her face up and laughed. "No," she admitted, "being able to click on a button doesn't mean that you know what you are doing. First you have to know what is going on behind the button."

"Thank you," said our teacher. "I'm glad you've been listening to what I've been saying in class when I've told you that we have to learn things the long way around before we can take the short cuts."

"Any other thoughts about what might be the hardest part of quantitative research."

I decided to go out on a limb and put my hand up. How daring of me. The teacher pointed at me. "Yes?"

"I think the hardest part is thinking about what kind of questions you want to ask, or rather thinking about the theories or hypotheses that you want to test."

"Why is that?"

Might as well go on, I thought, now that I had started. "Well, collecting data, as you said, just takes money and time. Interpreting the data, using statistics, means that you have to know a lot of maths. But, thinking about what to study and why, that needs some creativity."

My teacher looked impressed. I was also getting some raised eyebrows from the other students. I could see they were thinking about what I said.

"Good thought," said my teacher. "Do you agree?" the teacher asked the rest of the class.

There were lots of nods. Whew! I thought, relaxing a bit. I took a chance, gave an answer, and it didn't blow up in my face.

"I agree as well," said the teacher, "which is why I want you to focus on exactly that in our next exercise. Let me tell you what I want you to do. We're going to talk about counting people. At least, that's where we're going to start. Anyone know what that is called, when people are counted?"

"A census," answered Karl.

"Right," agreed my teacher. "Censuses have a long history. The word comes from the Latin, meaning to assess, but the Romans were not the first to have censuses. It is thought that the Babylonians were the first, in 4000 BC. They did a census so they could find out how much food would be needed to feed the population. The Romans did their censuses to find out how much tax each person should pay. Why does our country do a census?"

"To count how many people," said a few people at once.

"Just how many?" the teacher queried.

"No, how many, how old they are, things like that."

"Lots of things like that," agreed my teacher. "There are questions about each person in your household. Age, sex, marital status, where you were born, what languages you speak, where you work, what kind of job you have, how much money you make, how much education you have, what your house is like, if you've moved from a different house or location, if you have had a new baby, or a death in the family. Now, why do you think those kinds of things are being asked?"

We all thought for a minute and then Julie put up her hand. I waited. Julie usually made the class laugh.

"I don't think it is to see how much food we need," she said.

"Okay," laughed the teacher. "Not how much food, but maybe some other things we might need. What else could that be, Julie, since you have started us off."

Julie was funny but she was smart too. After a second she said. "I think knowing where kids are is important. Then you know where people will need schools."

"Great thought," said my teacher. "That is what is done by the governments, who do the census. Many countries do a census every ten years, but they are expensive and can be difficult to run, so not all countries manage to take a census very frequently. The governments want to know what the demographics of the population are. Demographics describes the structure of the population. Since the government wants to serve the population, they have to know how to serve them; what the population needs and where they need it to be."

"But there are other uses for census data too," continued the teacher. "Researchers use the data for quantitative research, as we talked about. And businesses use it too. Businesses want to know what kind of people live where, so that they will know where best to locate their businesses."

Max said, "Yeah, they don't want to serve the people, they want to make money off of us."

"And is that wrong?" asked the teacher.

"No," said Max. "It's smart."

The teacher laughed. "It works, let us say. They do make money. But how smart is it? Can you think of a reason that making money isn't always smart?"

I thought. I had a faint glimmer of something but it was hard to catch.

My teacher said, "You've all heard of Einstein?"

We nodded.

"A famous scientist. He worked with numbers and statistics and theories, but he also said this:"

> *Everything that can be counted does not necessarily count.*
> *Everything that counts cannot necessarily be counted.*

"What do you think he meant by that?"

The glimmer had turned into a flashing light. I put up my hand. The teacher nodded at me.

"I think Einstein meant that just counting where people are and things about them doesn't deal with what is really important in their lives. And I also think that making money from people isn't always the most important thing. We can buy lots of stuff, and make business owners rich, but having the stuff doesn't always make our lives better. In fact, it can make us sick, and the pollution that is caused when it is manufactured can make us even sicker."

"Interesting answer," said my teacher. "What do the rest of you think?"

"Like all the junk food that is so easy to buy and the stores that sell it being placed near schools. The businesses have it made, but it isn't good for us in the long run," said Jay.

"And the governments might see where people work and live," said Sian, "and build more roads so they can drive to work, but that is bad for the environment. So maybe census data isn't really any good for helping people in the long run, maybe it just solves problems for now."

"Another good answer," said the teacher. "But the census data produced depends on the questions being asked. If we weren't using the questions on the census that we have now, and that, as Sian has pointed out, perhaps do not address all our needs, do you have any ideas what kinds of questions we might be asking that would be better?"

I thought for a minute, but my brain was swirling. The rest of the class was quiet as well.

"Okay," said the teacher. "Let's take a break from our discussion while you do a little online research. I want you to look up a few websites, make some notes and then we'll go on with our work."

---

1. Look at these websites that view information gathering from a different perspective.

   While you visit the sites, write down answers to these questions:

   - Is this approach to gathering information useful? Why or why not?
   - What kind of theories could you test if you had this kind of information?
     - Gross National Happiness http://loc8.cc/otw/gnh
     - OECD Better Life Index http://www.oecdbetterlifeindex.org/
     - Happy Planet Index http://happyplanetindex.org/
     - Measure of Australian Progress http://loc8.cc/otw/ausstats
     - Genuine Progress Indicator http://loc8.cc/otw/rprogress

2. Look at these government census websites to see what information you can find about approaches to census and information sharing.

   While you visit the sites, write down answers to these questions:

   - What country's data do you think is easiest to access?
   - What datasets would you be most interested in?
   - Does any country collect something that others do not?"
     - http://loc8.cc/otw/census_ca
     - http://loc8.cc/otw/ons_census
     - https://www.census.gov/data.html
     - http://loc8.cc/otw/abs_census

I read, I made notes, I thought. Later my teacher called us all back together.

"I'm eager to hear your thoughts about what you have read, but I want you to do some more work first before we have our discussion. I'm going to hand out an assignment for you to work on. Read the requirements and then see what you can do. You'll have a week to work on this."

I took the paper that was passed to me and started to read.

The data collected during a census is input into tables. Since GIS has become prevalent, the tabular data has been displayed on maps so that it can be interpreted more easily. In addition, it gives the interpreters more ideas when they can see how the pieces of data relate geographically to each other, i.e. how data correlates.

Census data is sensitive and privacy issues are carefully considered. Although each household answers questions about itself, data are aggregated into census areas of varying sizes and hierarchical structures so the data collected about individuals is not revealed. Some governments let researchers, business people and individuals download the aggregated data. Since the data has to be usable by such a variety of people, it is not always in a format that is ready for the use you may have for it. It may require some work to make it ready. Usually there is a geographic data file for use in a GIS and tabular data that is in formats that are easily used by the majority of users, such as comma separated value format (CSV).

To join the tabular data to the geographic file requires a common field. You have to find the common field and join the table to the geographic file, the map. Sounds easy, but sometimes it is a bit convoluted. If you want to give it a try:

- Go to a country census website and look for their census data.
- Look for the geographic file (the shapefile).
- Look for tabular files and download them.
- In QGIS, add the shp file.
- Examine the names of the shapefile fields and the field contents.
- Open the tabular file in a spreadsheet.
- Examine the names of the spreadsheet fields and the field contents.
- If the field names are the same, that will be the one you need to use for a join. If they are not, check out the content of the fields. There will be one that matches in the table and the shp file.
- Click on *Properties* of the Layer in QGIS.
- Click on *Join Tables* and fill in the dialogue.
- Click *OK*.

To complete this chapter, you will create your own data as follows:

- You will create a spreadsheet with a field named `Join`. In that field, there will be rows with the name of the area, for example, A-Z. Have at least twenty areas.
- The columns will be numbers that reference questions, for example, 1–20. Have at least twenty questions.
- Make up questions you think will be interesting to work with—see the questions listed below. Your questions should reflect information you think is important to collect. Your perspectives will determine what that is.

- Fill in some numbers to reflect how you think people will answer.

Your table, with aggregated data reflecting percentages, might look something like this:

	1	2	3	4	5	6	7	8	9	10	11	12	13	14	15	16	17	18	19	20
A	20	33	85	10	45	20	50	10	90	55	45	22	70	8	70	80	80	8	55	8
B	10	22	88	32	21	10	56	9	80	75	43	10	75	6	73	89	85	6	45	67
C	22	10	67	50	78	8	67	5	32	88	42	9	56	5	67	74	90	2	33	4
D	42	37	76	14	32	7	68	7	45	89	33	6	80	8	62	32	56	10	67	56
E	7	56	76	28	80	8	45	9	83	76	38	2	84	9	88	36	80	8	80	7
F	2	12	67	32	67	3	15	20	6	22	67	48	78	10	82	22	77	9	45	43
G	4	32	69	26	32	17	19	16	7	11	69	21	58	20	56	10	54	10	48	45
H	19	9	90	18	17	20	28	7	18	9	78	17	59	22	83	69	59	10	33	89
I	20	4	78	11	67	11	32	8	27	7	22	9	67	70	73	63	60	10	26	42
J	5	58	92	8	50	8	88	22	82	4	21	7	80	74	90	50	49	12	73	67
K	11	43	38	62	32	9	64	28	45	90	18	6	50	78	85	52	90	9	78	5
L	42	22	55	45	76	4	57	20	79	67	19	20	69	63	74	59	92	8	34	8
M	6	24	72	23	11	17	32	15	45	45	22	11	74	50	79	48	38	20	38	32
N	22	21	84	16	86	17	39	30	74	53	55	10	72	51	33	89	96	5	48	22
O	31	8	83	19	64	4	17	38	89	62	59	6	62	6	39	73	78	9	51	34
P	12	5	82	18	53	21	28	18	22	32	69	8	69	8	38	76	79	10	48	56
Q	2	34	92	8	26	29	63	11	16	38	78	20	89	4	35	10	84	11	82	78
R	24	36	57	43	48	31	78	19	8	73	88	26	83	34	69	28	83	18	64	42
S	33	61	86	26	77	4	82	9	24	72	32	32	56	38	78	39	52	24	69	55
T	37	25	61	39	31	7	29	20	7	19	38	30	67	48	48	49	45	26	22	62
U	16	27	71	31	68	10	17	18	10	20	29	21	69	42	92	78	89	12	29	67
V	10	31	95	5	52	32	37	16	36	32	18	22	78	55	64	73	67	15	39	3
W	28	38	42	59	38	6	38	9	48	54	89	9	73	59	60	69	69	8	41	2
X	33	41	68	42	51	50	42	8	28	48	55	4	53	7	52	65	90	9	72	78
Y	18	9	69	32	27	3	44	21	42	38	67	33	44	4	31	64	78	11	67	67
Z	10	4	74	30	79	7	50	17	80	72	43	29	50	9	38	50	52	20	60	8

Note that all questions need answers that are numerical. Your census will have specified that answers will be 1 if yes and 0 if no. Data from all families will have been aggregated and percentages are reported in the table above.

Here are some sample questions:

1. Are there children under 10 years of age in your home?

2. Are there people between 10 and 25 years of age in your home?

3. Are there people between 25 and 60 years of age in your home?

4. Are their people over 60 years of age in your home?

5. Is your family happy?

6. Are there people who did not complete secondary school?

7. Are there people with an undergraduate degree as their highest level of education?

8. Are there people with a graduate degree?

9. Does your family commute more than 10 hours to work in a week?

10. Does your family eat at least one meal a day together?

11. Does your family feel financially secure?

12. Would your family relocate if they could?

13. Do you think your family is physically healthy?

14. Does your family have enough green space within walking distance of home?

15. Does your family feel safe in their neighbourhood?

16. Does your family watch more than 10 hours of television in a week?

17. Does your family spend more than 10 hours online in a week while at home?

18. Does your family read or play games together?

19. Does your family feel supported by their government?

20. Is your community important to your family?

- Type your data into a spreadsheet table with 26 Zones in rows and 20 questions in columns as in the table above. Save your table as a CSV file.

- Create a geographic file in QGIS, a shapefile, with one large polygon and a field named `Join`. Then use the *Split Feature* tool to create 26 areas. Populate the field called `Join` with the letters A to Z (Figure 6.1, on the next page).

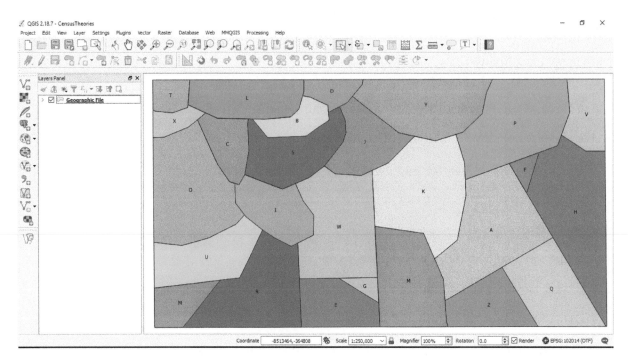

Figure 6.1:

- Give your table to one student in your class and your questions to another. In turn, you will receive a table and questions from two other students.

- Add your CSV file to the QGIS GUI through *Layer->Add Layer->Add Delimited Text Layer* (Figure 6.2).

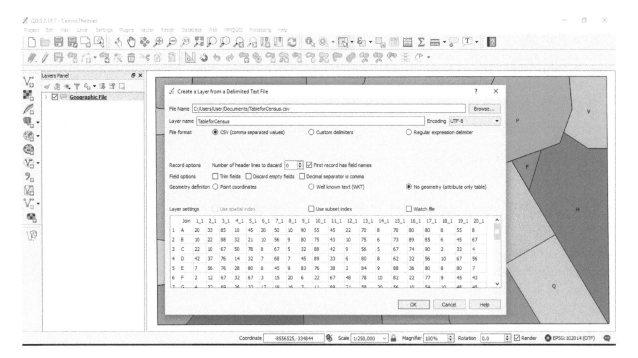

Figure 6.2:

- Join your table to your shapefile by right clicking on the Geographic File and selecting *Properties->Joins*. Read about Joining tables in the *User Guide, The Vector Properties Dialogue* and in the *Training Manual, Module: Forestry Application.* See Figures 6.3 and 6.4).

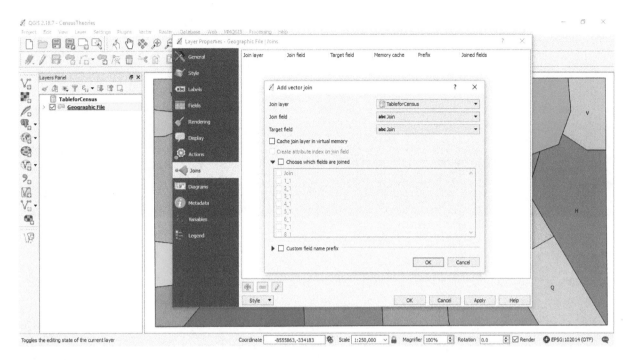

Figure 6.3:

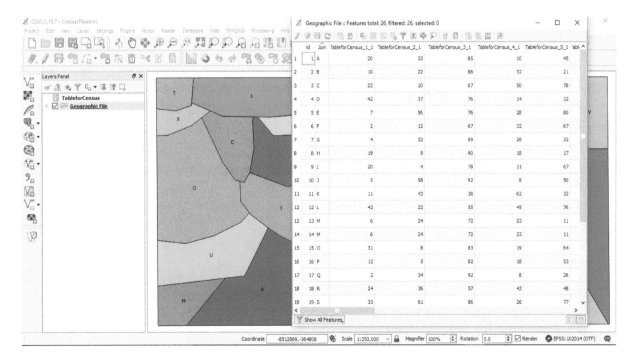

Figure 6.4:

- Write five hypotheses about your data. Hypotheses based on the questions in this example, might be: "Families who have green space near them are more likely to be healthy" or "Families who watch less than 10 hours of television a week are more likely to be happy."

- Do Basic Statistics on five fields (questions) using *Vector->Analysis Tools->Basic statistics for numeric fields* (Figure 6.5).

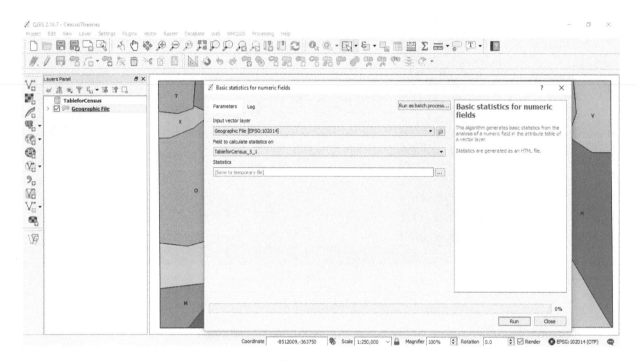

Figure 6.5:

An elected official might want to know, for example, if their constituents are (question 5) happy, (question 15) feel safe and (question 19) feel supported by their government, on average. Record the mean for each of the five questions you have selected.

- Choose five different questions and display the data with classifications. Review how to classify data in the *User Guide, the Graduated Renderer*. Figures 6.6, on the facing page and 6.7, on the next page illustrate Natural Breaks, 3 classes, for Question 14, regarding green space.

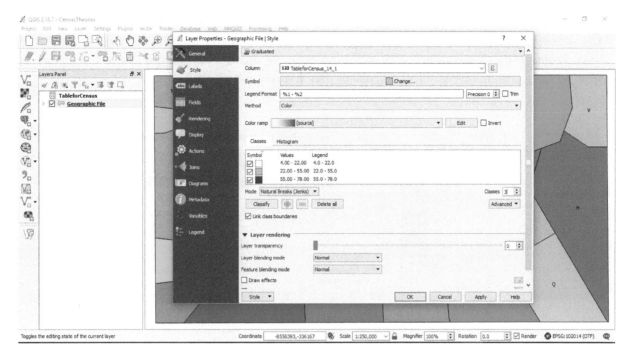

Figure 6.6:

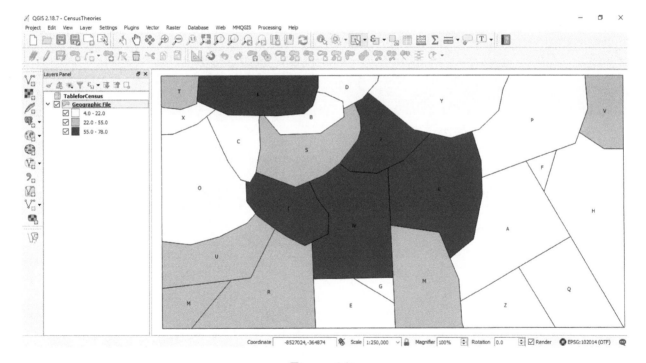

Figure 6.7:

- Write up some conclusions about what you observe from the classifications. You will have to be creative in working with data that is purely fictitious, but that will give you more opportunity to stretch and develop your creative brain cells! What can you invent?

- Investigate and illustrate your hypotheses by using the *Select Features using an Expression* icon from the *Attribute Table*. Read about selecting features using an expression in the *User Guide, Working with Vector Data, Expressions*. The following example illustrates: "Families who have green space near to them are more likely to be healthy." See Figures 6.8 and 6.9.

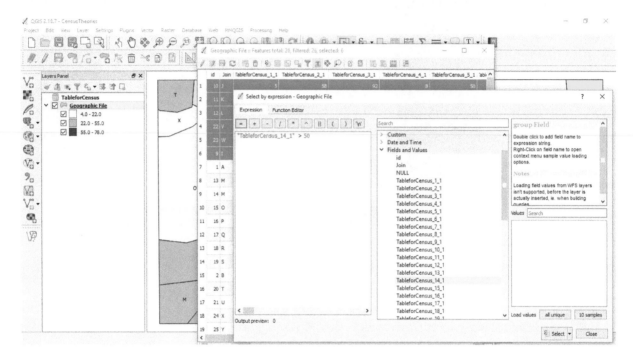

Figure 6.8:

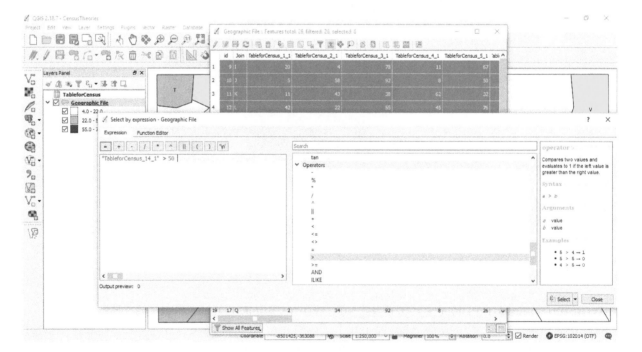

Figure 6.9:

The results from the selection in Figure 6.8 show that 6 areas have 50% reporting they have enough green space and Figure 6.9, on the preceding page shows that 5 of the 6 areas think their families are healthy. Thus, the hypothesis MAY be correct since there appears to be a correlation. However, there are many reasons why this hypothesis would need considerably more checking. Can you think why? List at least two reasons. The areas of correlation are highlighted in yellow on the map (Figure 6.10).

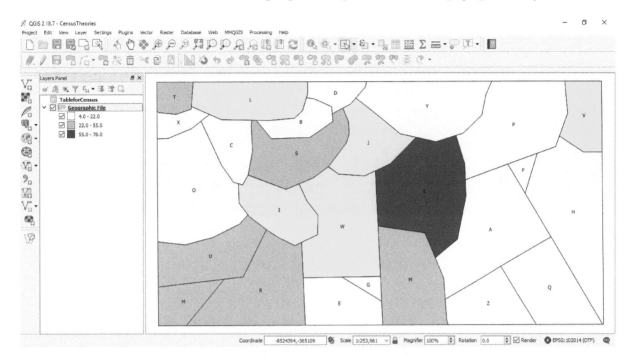

Figure 6.10:

Note: If you have chosen to download and use real census data, follow the same steps as for the invented data. Create hypotheses, test them, speculate on your results, map the data to illustrate your conclusions and evaluate your interpretations.

7. Rooting for Web Mapping

Learning Tip Seven: If you care about something, do you work harder? Does caring make you want to do your best? What makes you care? If you always wanted to do your best, how could you approach your work? How could you inspire others to care? Is it ever okay to just be good enough? Why or why not?

"Holly. Holly! Are you listening? Holly!"

"Mmmm, yes," Holly moved a bit. "Hi, Iris, what's up?" she asked.

"Wake up Rose, Holly. I want to talk to both of you."

"Iris," protested Holly, "you know Rose is tired these days. She's delicate. She's...you know...old."

"I heard that, Holly, you young pip-squeak. I am not old. I am..." Rose searched for a word. "I am dignified, stately, distinguished," she flounced, in annoyance. "And even if I am past my first blush of youth, at least think of a better word to describe me. *Old* does not sound becoming at all. Anyway, I do appreciate your concern for me but I have been awake for hours, watching. The forest is beautiful this time of the morning. It is ever so peaceful. Listen: the sighing of the wind in the trees, the rustle of the little creatures running under the ferns, the whisper of the water flowing to the ocean. I do love morning. Now, what was it you wanted, Iris, dear."

"I wanted to tell you both some exciting news", said Iris, thrilled to have Rose's attention. Rose was such an epitome of beauty, Iris thought. She would have used that to describe Rose, not "old", but she didn't want Holly to feel that she was using fancy phrases to make her feel uncomfortable. Holly was a good, loyal friend, even if she was sometimes a bit prickly.

"What's so important that you need us both at this time of day?" asked Holly. "Do you need the others, too?" she asked.

"Not right now," said Iris.

"Sorry, Iris," Rowan said, quietly. "I couldn't help but hear. I don't mean to intrude."

"Rowan," Iris said, "you are always welcome. Don't worry. I would have told you soon anyway. I didn't know you were so close that you could hear me."

"I'm growing again," said Rowan proudly. "I can hear so many more things now that my roots are spreading."

Iris felt a bit sad that her roots were so small and she could only reach to a very few others. She had to get Holly to relay her messages to almost everyone in the glade. Still, she thought, I may be small, but I have important news.

"Everyone, listen," she began again. "I have important news about something called the World Wide Web. Robin told me," she paused for breath and was interrupted immediately.

"Are you talking to spiders, dear," asked Rose. "Spiders are lovely. Those webs, filled with dew in the morning; in the morning light, sparkling like jewels," Rose mused, sighing her appreciation of her spider friends.

"Don't be silly, Rose," interjected Holly. "She's been talking to birds. What robin have you been talking to, Iris. I know you like those little birds, but frankly, I think they have beady little eyes. I wouldn't trust a robin."

"I don't mean spiders or robins," Iris said, a little flustered that the conversation was taking such quick turns. She could hardly keep up and she wasn't getting her message across very well.

"Everybody," said Rowan. "Let's give Iris a chance to explain. Iris," he said, "please go on. What is the world wide web and who have you been talking to?"

"Thank you, Rowan," Iris gave his long, thin root a squeeze with her short, stubby little roots. "Robin is a human. You know, the ones that walk by on the path."

"Oooh," shuddered Rhodo. "Sorry, couldn't help but hear what was going on. I don't like those humans. Have you heard about my family? Have you heard what they are doing to us?"

The group had, indeed, heard, in fact, they had heard on a daily basis about Rhodo's family tragedy. However, they knew the conversation would not continue until they heard it again. So they braced themselves and waited for the inevitable.

Rhodo didn't wait for an answer, she started right in, "My family are being eradicated, terminated, ripped out by their roots," she cried. "And why?", still she did not wait for an answer, "because they are a non-native species. Did we ask to be taken from our place of origin?" she asked. "No, we did not. Did we ask to be planted here? No, we did not. Can we help it if we do well here? No, we cannot", she stopped ranting and finished with a half sob, "And people say they love us, they say we are beautiful. You think so don't you," she implored her friends. "We don't mean any harm..." She dropped petals at an alarming rate from her brilliant flowers. A carpet of crimson spread beneath her leaves.

"Rhodo," soothed Iris, "you know we love you. And we are so sorry to hear about what is happening to your family. I know that since you heard about it, since you felt it through the network, that you have been distraught with grief. We feel for you." The others agreed and tried to comfort Rhodo, each in their own way.

"Not everyone hates you," Rowan soothed. "Don't forget, lots of people plant rhododendrons in their gardens because they love them so much. People do love us, all of us."

"Sure they do," snipped Holly. "As long as they can keep us cut back and contained. As long as we are some use. Berries, fruit, flowers or nice leaves, as long as we produce, we are 'allowed to live'. Do they care about us otherwise, I think not!"

"Iris, tell us about your important news," Rose instructed, in a soft way, while she sent comfort to Rhodo who was still shivering and shaking. She leaned close to Rhodo and rubbed leaves, tickling her a little to see if she could make her laugh.

"Okay," Iris started again, hoping this time she could continue. "Robin is a person, not a bird. She walks by here all the time. Yesterday she stopped and came over to me. I thought she was just going to sniff, but she sat down and," Iris shivered a bit with the remembrance, "she touched my petals."

"Happens to me all the time, dear," said Rose.

"Not to me," said Holly. "My leaves keep those creatures back," she said with pride.

"And not only did she touch me," said Iris quickly, hoping she wouldn't lose her audience yet again, "she talked to me and she cried."

"Cried?"

"Yes, a tear fell from her eye onto my petals. It was salty. There was lots of information in it."

"What kind of information?" asked Rowan.

"About the World Wide Web," answered Iris. "Robin also called it the Internet. She said that she was so sorry they were going to kill us all."

"What?", everyone shrieked. "What are you talking about? Kill us? How? Kill us? Why? No one has told us," they said.

"Listen," Iris started again. "I know you haven't heard about this. I only just heard, myself, and I can tell you more if you will just listen. Robin said that lots of people like to get to the ocean from here, so many people that they don't have enough space to park their cars. You know, the noisy, smelly things that they sit in to get here. They want more parking space so they are going to cut us down and pave us over." Iris stopped for a moment to allow them to absorb the information.

"But," she continued, "Robin cares about us. She read things about us. People are starting to understand us."

"Sure," said Rhodo, "they like plants and trees, just as I said, if they are the right kind and if they grow in the right places and in the right ways. They don't want us to give too much shade or too little shade. They don't want us to drop our leaves or seeds around. I don't call that understanding."

"No," agreed Iris, "that is not understanding. Robin said that they do understand that we benefit them, not just for our beauty and for food, but for flood control and for cleaning the air. But they are starting to understand more about us, about our feelings."

"How can they know that, dear?" asked Rose.

"They have done studies that show that we communicate with each other through our roots. They have other studies that show we react when there is heavy traffic in areas. They are starting to see that we are sentient beings just as they are."

"Well," sniffed Holly, "I'll believe that when I see it. I bet we feel our roots being ripped out before that happens."

"We may," said Iris. "And we will have to face that if it happens." The others felt her quiver. "But Robin is going to try to help us with the World Wide Web, the Internet. She is going to create a web map that will show where we are and who we are. She is going to try to tell as many people as she can that they don't need a bigger parking lot, what they need is to change their ways. She wants to tell them to come to visit us without cars, to come on foot or on bicycles, you know, the things with wheels that don't smell and aren't loud."

"Do you think it will work?" asked Rowan.

"I don't know," answered Iris, "but we can help her and maybe, even if we lose, we can help others who may be facing the same fate. All we can do is try our best."

"Thank you, dear," said Rose. "You've given us hope. Let's tell the others now. Everyone needs to know."

Want to try to help the glade? Making maps is a powerful way to illustrate information and making a web map allows that information to be shared with a lot of people.

To complete this chapter, you will follow the next three steps, numbered below. You will make a map of the glade, its inhabitants and the surroundings. You will need to show the parking lot, the access roads/trails to the parking lot and the walking path to the ocean. You will also use the map to achieve the following:

- Influencing travel by illustrating pedestrian and bike friendly routes
- Developing interest in the glade by mapping plants and trees
- Improving awareness of other natural areas by linking or pointing to them

After you make the map, you will view it in a browser and you may publish it to the Internet so that you can share it with anyone and everyone.

1. First, as always, you should do some reading and investigating about the plugin or plugins you will install.

The first plugin is `qgis2web`. This plugin will allow you to prepare and create a web map. There are a multitude of settings for you to choose in this plugin. Having read about it, using the links below, you will be able to make your web map the best it can be. After you export your map, you can view your map on a browser on your PC but it will not be shared with others until you publish it to the Internet. For *Export to Folder*, choose a file location that you can access. Using the HTML file named `Index` from that exported folder will allow you to open the map when you want to revisit it.

`http://loc8.cc/otw/qgis2web_tutorial`

`http://loc8.cc/otw/qgis2web`

The qgis2web tutorial is also available in the data package, available at: `http://locatepress.com/otw_files/otw_data.zip`.

The second plugin is the `QGIS Cloud` Plugin. You may or may not want to use this plugin, depending on if you want your map to be on the Internet for others to see. To publish to the Internet, you will need to create a free account.

http://demo.qgis.org/

https://qgiscloud.com/

The video link below is a bit lengthy, but will walk you through the entire process of creating a web map.

https://youtu.be/4gyW1aeoTvU

2. Follow along with these steps to prepare a basic web map. This basic map will give you practice for Step 3.

Create a shapefile for points with fields that you can use to categorise your data (Figure 7.1).

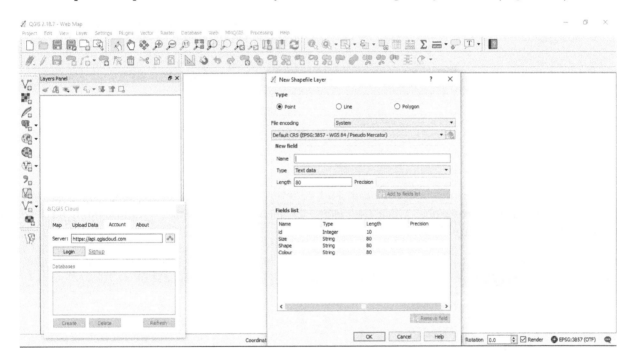

Figure 7.1:

Save the shapefile and edit it to add points and to populate the attribute table (Figure 7.2, on the following page).

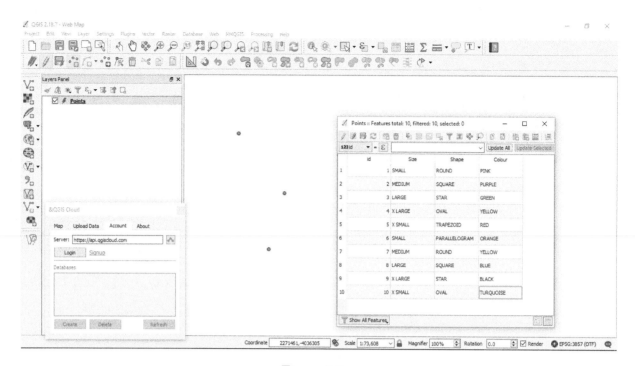

Figure 7.2:

The id field is useful for working in QGIS, but isn't of interest to the user of the map. To hide it, right click on the Layer, select *Properties*. Click on *Fields*. Click on *Text Edit*, under *Edit widget*, next to the id field. Select *Hidden* (Figure 7.3).

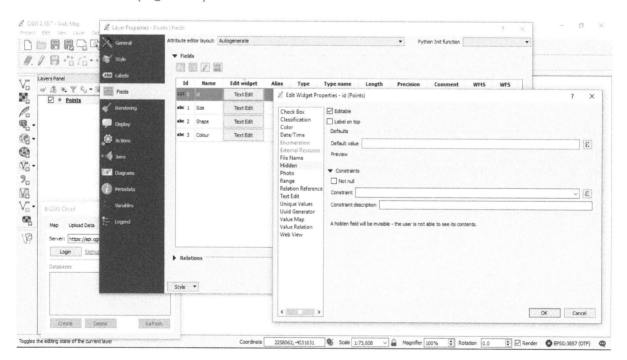

Figure 7.3:

Click *OK* and *OK* to close the windows. Using the *Identify* tool in the map and under *Actions*, you will see what the map user will see.

Under the *Style* tab, categorise your map to make it more interesting for the user (Figure 7.4).

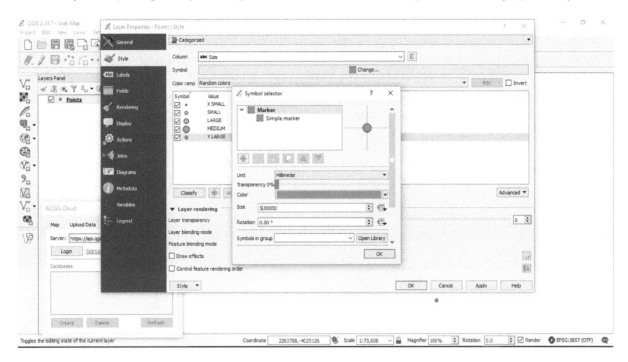

Figure 7.4:

To create the web map, click on *Install and Manage Plugins* and install the `qgis2web` plugin. Access the plugin from the *Web* Menu item. Click on *Create Web Map* (Figure 7.5).

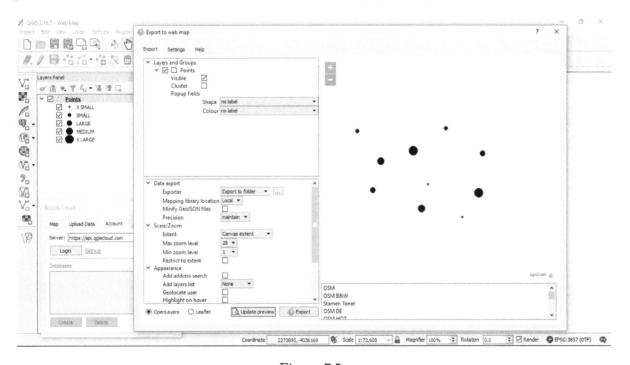

Figure 7.5:

Click on *Export*. A browser will open. Click on one of the points and a pop up window will show the details about the point.

3. Prepare a detailed, persuasive web map and publish it, as per the instructions for completing the chapter, above. How can you make your map detailed and persuasive? Read about what you can do in the help files. Investigate all of the parameters that can be used in the Export interface. Know what would make you care and see if you can add it to your map to make others care as well. Look at some other websites for designs. Inspire your audience to save the plants and trees with the information on your map!

8. The Ins, Outs and Overs of Working in GIS

Learning Tip Eight: Experimenting with your own, created data is a great way to learn how tools work. Read about the tool in the help file. Create, use the tool, observe the outcome. Review the help file. Taking the time to read and experiment for yourself is never a waste of time and will prevent that frustrating feeling you get when software doesn't behave as you expect it to.

My eyelashes were frozen together. I couldn't see a thing. There wouldn't have been much to see anyway since it was so dark. Even at noon, when I stopped the snowmobile to eat my cold (make that frozen solid) sandwich for lunch, there was barely a hint of light. Most of the light was from the moon shining over the bent forms of stunted trees. The moon? At noon? The Arctic is a strange place!

I took off my mittens and used my cold fingers to melt the ice on my lashes, rammed my hands back into my mittens and vowed to keep my eyes blinking more quickly so that it wouldn't happen again. But what could I do for my nose and lungs? The freezing air, even filtered through my balaclava, felt like it was stabbing me on the inside. This was serious cold. Who would be out on a day like this, I thought? I never thought it would be me.

So, how did I get here, you might ask. Well, back in the office where we are working on a project to determine the best site for a new wilderness area, a sanctuary for caribou, I am a senior person in the office. That is, I am not the newest person in the office. The newest person only started last week. I started six months ago, at the beginning of the summer. A veritable veteran, I had survived the Arctic summer!

During the summer, all of the data for the areas that were being investigated for the sanctuary had been captured by helicopter and plane, with cameras mounted to the body of the aircraft. We could have used satellite imagery as the resolution was getting much better, but the best was out of range of our budget. We wanted a resolution of 40 cm with colour imagery which meant that anything bigger than 40 cm could be seen in the photo. Black and white imagery was cheaper, but colour would be easier for everyone to understand. We had to sell the area to decision makers so that the caribou would be protected. Decision makers were not necessarily experts at interpreting aerial photos. Colour might make it easier. Was colour an unnecessary expense? Maybe, but we didn't want to take any chances with losing the opportunity for the best caribou protection.

Now that the photos had been taken, my boss needed someone to go out, take photos from the ground, and ground check the areas that had been flown, to see if there were any surprises and to make sure the interpretation work had been done correctly.

In the summer, I had gone out by helicopter, landed, well, more like hovered in case the helicopter got mired when it landed in permafrost. It would not have been a good thing to be hundreds of kilometres from civilization with no way to get out. It would not have been a good thing to have to dig a helicopter out of spongy, wet ground. But the worst thing was that it would have prolonged the exposure to mosquitos. I am sure you have all experienced being bitten by mosquitos, but magnify that by 1,000, maybe 10,000, because that is how it feels to be with the mosquitos in the Arctic in the summer. Most people don't know how hot it can be, but this is continental climate: hot, humid summers and cold, dry winters. It can be over 35 degrees Celsius and humid in the summer. The choice is dying of the heat wearing a hoodie, bandana, gloves, long sleeved shirt and trousers tucked into socks, or being eaten alive? Let's just say I lost a lot of water through sweat and I still did not come away unscathed and unbitten. Mosquitos in the Arctic don't mess around. They are in a hurry: a short season and a lack of victims turns them into aerial piranhas.

But in the winter, it was cheaper to eliminate the cost of a helicopter and do photos and ground checking using a snowmobile. Before I started, I hoped I would be able to finish the outdoor work in the summer, using the helicopter. After the mosquito attacks, I was sure I would prefer doing the work in the winter—or letting someone else do the work.

"What about the new person," I asked my boss.

He looked over at the new person, sitting, hunched at the computer. "Are you kidding?" he asked me. "Ninety pounds, wet. I can't send person like that out. What if something went wrong? Not enough meat on the bones to survive for a day." He looked me over. "Now you on the other hand...," he chuckled, punching me lightly on my slightly larger waistline, "No worries about you making it through the night, waiting for a rescue."

"Thanks," I said, with a growl.

"Got your radio?" he asked.

"Yes," I answered. "Safety first."

"And the camera?"

As if I would forget the reason I was going out. "Yes," I answered, "and I'll take lots of pictures although I don't know why. The aerials have all been great so far."

"I know," he said, "but our contract requires ground shots for verification and context. Have to follow the rules. You know," he said, "we might be doing this for the money, for the work, but we do have to remember that we also want to do the best job possible. The caribou are counting on us."

He was right, he cared, I cared, lots of people cared that the caribou would have a safe sanctuary. Despite the Arctic being huge, the places that were left for the caribou, the places they really needed to be, were declining in size. The best of what was left needed to be mapped, analysed, and protected with boundaries. I was part of an important project.

Still, eating my rock-hard sandwich, standing by the snowmobile, blinking as fast as I could to stop my eyelashes from freezing together again, I thought about the belief that winter work would be better than summer work. No mosquitos, it was true, but this cold was unbelievable. I gnawed off as much as I could of the now-frozen bread, sighed and awkwardly took the camera out of my pocket to take a shot of the area.

Oh great, I thought, clicking on the button to turn it on. The camera is frozen again. Oh no, I thought, I'll have to warm it up.

Warming up the camera meant that I would have to put it inside my parka. Not against the skin this time, I reminded myself. Yikes, that had been cold. This time I slid it between my first and second (of five) layer of clothes. Still cold, but it didn't burn the skin. I waited, humming to keep myself cheerful.

I looked around at the white ground, at the black sky. Humming sounded strange in the absolute silence of the wind-still day. After a few minutes, I took the camera out and looked for the best possible angles. Quickly, mind, before it froze again. I snapped a few shots, got back on the snowmobile and started back. Mosquito plagued summers versus freezing cold. At least they didn't come at the same time. I looked forward to getting to the office where it was warm and mosquito free, as well.

When I arrived, I downloaded the photos from the camera, confident that all was as it should be with the aerial images. I started working on mapping and analysing the data. Before long I was wishing I had earplugs, it was so noisy, with all the people talking on the phone and discussing strategies at their desks. Then the person on one side of me started eating a tuna fish sandwich. It was pungent. What an assault on my nose! The person on the other side was crunching on crisps. I glanced over. Just my luck, it was an extra-large bag. Was office work really better than field work? You tell me. Oh well, I had to get to work on the map.

I had the aerial images. From them I was going to outline areas for both summer and winter feeding grounds. Caribou food is lichen. In the summer, caribou want to be in the wind so they get relief from the mosquitos. I could sympathise with that. They eat lichen growing on the ground. In the winter, they want cover from the wind to help them preserve body heat. I could understand that as well. I shivered a little thinking of my frozen adventures.

"Finished yet?" my boss interrupted my thoughts.

"I haven't even started!" I protested.

"Kidding, only kidding. But that's great. Since you haven't started, I have a job that only you can do. And it starts right away. Well, as soon as you can delegate the mapping to the new person."

I had a bad feeling that my new assignment meant I would be putting on my winter gear again. Could I protest?

"I'm not sure the new person knows enough to do the mapping."

"Great then, no time like the present to learn. I'm sure you have some notes from when I showed you how to work with the software. Share the notes, explain the project. Meet you in a hour outside."

Somehow, I was not surprised. I sighed, dug for some papers in my desk drawer and went to the new person's desk.

"Hi," I began. "I've got a job for you. The boss wants me to give you the instructions. I have an hour and then I have to be off. Luckily I have some notes written down of the way I learned how to do the work so you should be fine."

The new person looked worried and I could sympathise, but my boss was right, jumping right in was the best way to learn.

"Doing what?" asked the new person.

"You're going to create a map with proposed areas for a caribou sanctuary. You are going to decide potential areas that will go in the sanctuary and areas that will be out and you're going to do it using overlay analysis. Think: In, Out and Over," I laughed.

"First, I suggest you study up a bit," I said. "Use this list to get started. Reading about what you will do is a nice introduction before you get started. It may seem too much to take in all at once, but it paves the way for the learning you will do when you start clicking on the program. And it provides you with the places you will need to go when you run into difficulties. Without the foundation, you are likely to bang your head against the program for a long time. Take my advice: read first."

- *User Guide, QGIS Processing Framework, Vector menu, Geoprocessing tools* are core to the concept of GIS.

- More QGIS documentation can be found at `http://loc8.cc/otw/qgis_intro`

- `http://loc8.cc/otw/buffers`

- Look online for YouTube videos by searching for QGIS Geoprocessing.

"Second, I suggest you test out geoprocessing tools by creating a new project in QGIS. When you use the tools, you will be able to see how they function. Have a look at this graphic (Figure 8.1) and make one just like it. In QGIS, go to **Vector->Geoprocessing** and know how to work with the Overlay Analysis tools: Intersection, Union, Dissolve, Buffer, Difference and Clip. I know you know how to create shapefiles and how to symbolise them, but just in case, you can always look at the helpfiles for review."

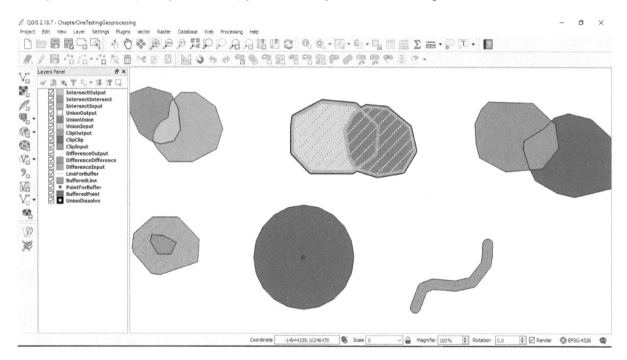

Figure 8.1:

For example, let me show you how I would work with Difference.

1. Look at the tool. It needs `Difference input`, `Difference difference` and `Difference output`. So you need three shapefiles (Figure 8.2, on the next page).

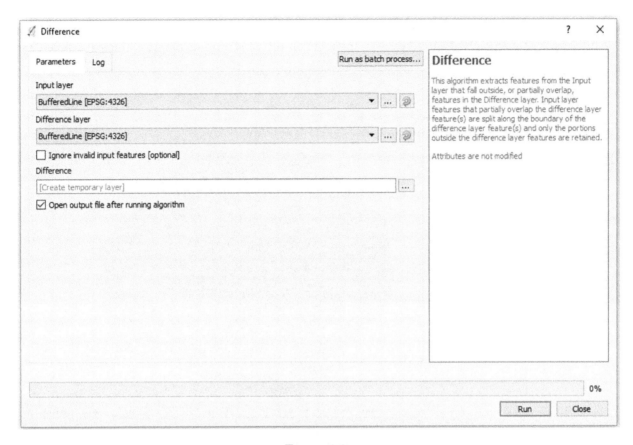

Figure 8.2:

Create a shapefile for Difference input (Figure 8.3).

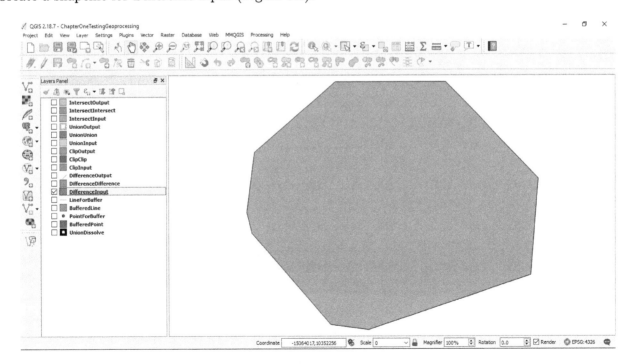

Figure 8.3:

Reading the help file associated with the tool tells us that the tool works by taking out what is overlapped by the difference shapefile. If the difference shapefile is on top of the input file, it will leave a hole (Figure 8.4).

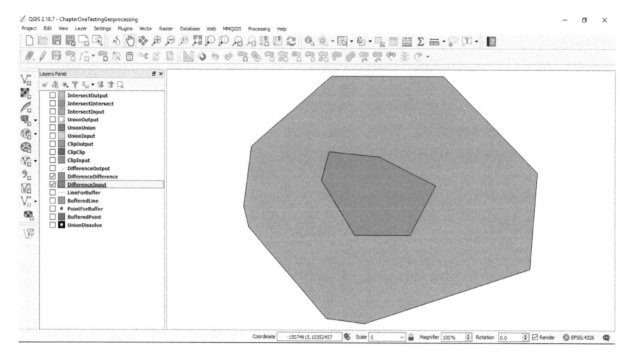

Figure 8.4:

Once the tool has been run, the output file shows the hole (Figure 8.5).

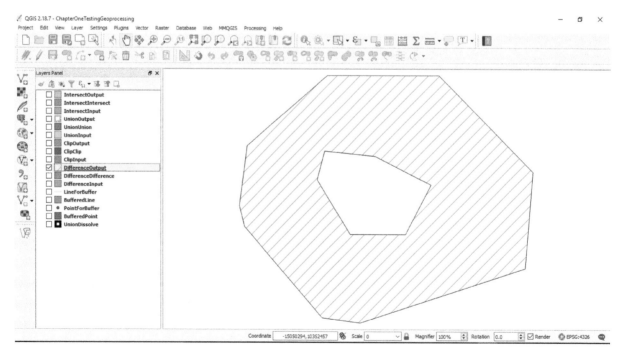

Figure 8.5:

And, if I use the right kind of symbology, I can illustrate the process when I show all three layers at once (Figure 8.6).

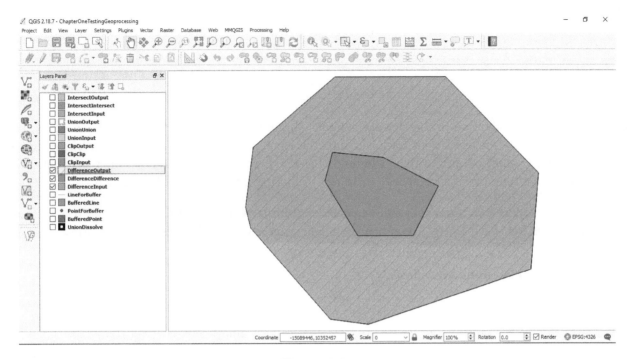

Figure 8.6:

"You'll do the same for all of the Overlay Analysis tools," I told him.

"Third, have a look at this (Figure 8.7). It's the area where you need to be for creating the map."

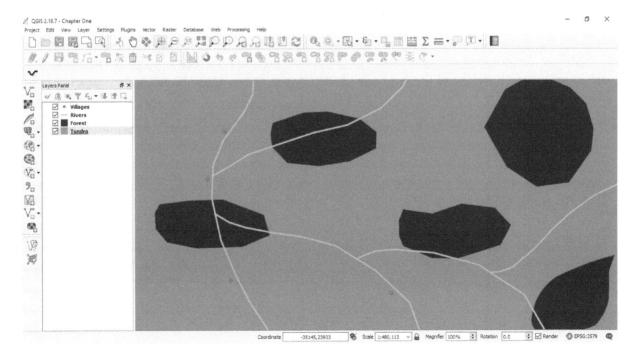

Figure 8.7:

Figure, on the previous page illustrates a potential spatial arrangement for the area, but you can be creative about a layout of forested and tundra areas as well as river and village locations. Caribou require a large range. Use an appropriate CRS from Options: CRS [a] 1 and a small scale, as is shown.

[a]Settings/Options/CRS (Coordinate Reference System). Where do caribou live? Yukon and Alaska are two likely locations. Search for, and choose, a projected CRS using the globe icon on the right of the CRS interface. A small-scale map will be, roughly, over 1: 250,000. The scale and coordinate system are displayed on the tool bar at the bottom of the map panel.

I continued, a bit tersely, due to the time, which was slipping by. I needed to explain thoroughly, but I also needed some food before I went back out into the elements. Being cold was tough, but being cold and being hungry would be unbearable. I figured the new person could handle succinct instructions. A little floundering around was the best way to learn. I didn't want to handhold too much.

"The goal is to produce two areas that can be considered for a caribou sanctuary and then to recommend one, with reasons for your choice. A good caribou sanctuary has the following characteristics:

- More than 25 km from any village

- Access to water

- About half of the area should be forested for winter cover and lichen covered trees

- The other half will be tundra for summer calving grounds and ground lichen."

The new person was flipping through pages and looking at my illustrations, as well as shooting me looks of horror. "I'm not sure I quite get...," he wrinkled his forehead. "I don't know what order to do things in."

"I can understand that. You'll need to plan and think carefully. You'll need to experiment a bit. You'll need to make notes and diagrams of the process. Analysis isn't easy, but it is fun because it's so challenging. And, no worries, you'll be fine," I said with conviction. "Just work your way through it. You'll be amazed what you can do. Now, some other things you need to know. Some ideas for you to get started.

- It is easiest to draw a large rectangle of tundra and put forested areas and rivers over the top. However, this will result in areas being combinations of tundra, forest and/or river which cannot occur. You could digitize the areas and ensure that the polygons are perfectly snapped at their edges, or you can use the correct geoprocessing tool so that areas cannot be more than one category. The latter option is the easiest. If you do not do so, your final calculated area results will be incorrect. As a challenge, I'll let you figure out what tool that is—don't you love challenges?

- A GIS always stores the location and area of features, but this information is not always in the Attribute Table. You will have to use the Field Calculator to display the area to show that the proposed sanctuary meets the required needs. Search for $area in the Training Guide."

The new person smiled weakly. "Okay. I'll do my best. What do you need as output and when?"

"I don't know when the boss wants it, likely you have a week, but I'll check to be sure. When you are finished, you will need to present this." I handed over the list.

The new person read:

- A map or screen captures of geoprocessing functions so that the stakeholders involved in the process can understand how the work has been done. Stakeholders always like a bit of background. Create a project similar to Figure 8.1, on page 92. The symbology will help to explain how the tools work.

- A map and insert table showing the calculated area for two different proposed sanctuaries. See User Guide, Print Composer, Chapter 19 and The Attribute Table Item, 19.2.6.

- Create a description of the process you followed, either written or as a flow chart so that you can explain how the work was done.

Your job is to create the work that the new person needs to produce.

"Listen, I have to go," I said. "Do your best. Maybe next time I'll get to do the mapping and you can go on the expedition. A word of advice: start stock piling warm clothes and start eating. You need a blubber layer to make it through the winters up here!"

9. Transformations

Learning Tip Nine: Being able to see new solutions, or to look at problems in a new way can change your way of thinking. When learning causes you to change your outlook, it transforms your view of the world. It's like changing levels in a computer game. A whole new world is there to explore. Seeking transformational learning can make your life very exciting.

"I'm in a rut."

I pulled my phone away from my ear and looked at it as though it would have an answer to my unspoken question. How could my friend be in a rut again, so soon?

Okay, I thought, I'll ask. "How can you be in a rut again so soon? Didn't I just pull you out of one?"

"You did," he answered with a sigh, "and I'm grateful for it although still slightly bruised from the experience, but it's happened again."

"Are you telling me because you want my help?"

"Yes. No. Maybe. I don't know," he admitted.

"You know what I think?"

"You're going to tell me anyway, aren't you?" he said.

"Can't resist," I laughed. "But it's short."

"Okay, tell me."

"A word to the wise."

"That is short. What does it mean?"

"It means, my friend, that if I tell you one word, and you are wise, it will be enough to change your life."

"So, what's the word?"

"Learn."

"Great. Thanks so much. That will surely change my life," my friend said, his voice dripping with sarcasm.

"I know you, don't forget," I said, being blunt but trying to be kind. "I helped you with your last rut and it changed your life, but it was at my suggestion that you made the changes. I don't think you really learned about changing things up for yourself. The learning you experienced was not transformational."

"Transformational? Great. Now you don't just want me to do more learning, you want me to transform."

"Exactly," I said.

"Arghh!" my friend exploded. "I don't think I'm ready for your word or words just yet. I'll talk to you later." He hung up.

You know, I think he is right. When he's ready, a word is all that it will take. I put down my phone and got back to my own work. He'd phone again if he needed me.

As it turned out, I didn't have long to wait. Exactly 24 hours as it turned out.

I had just made myself a pot of tea and was sitting down at my laptop when my phone rang again.

The conversation began in much the same way as the previous day's call.

"I'm just so frustrated," said my friend. We knew each other so well that he didn't even bother with the preliminaries such as "How are you?" or "What are you up to?" No, he jumped right back into the fray, so I accommodated him and asked, once more.

"Why? I thought things were going so well."

"I guess in some ways. I mean, since your last suggestions, my life has changed so dramatically."

"And I thought you were happy."

"I am. I know I must sound like a whiner. Listen, can you come over. I want to talk more about what you said yesterday."

"Sure," I said, looking at my pot of tea and my unfinished work. But, the tea could wait and so could my work. I had a friend in need.

Half an hour later, I knocked on my friend's door and then walked in as I usually did. Having been best friends for years, we treated each other's houses as our own. He was sitting at his dining room table, just off the front door.

"Look at me," he said. "I'm a total mess."

"Well," I admitted, wondering how truthful I could be without hurting my best friend's delicate feelings, "it does seem as if you should be happier than you look. Your face could rival a thunder cloud for first place in menacing appearance."

"Look at me," he said again, disconsolate. "Dog at my feet. Actually, dog on my feet. Cutting off my circulation. Cat on my lap. You know, if you hadn't let yourself in, I would have had to move and I haven't moved for 2 hours. My behind has turned to cement."

"I guess the pet weights you're wearing don't help. Are they to help with heating costs?" I joked.

"No, they just think that I'm so static when I work that they might as well settle in. Sometimes I feel as if they settle in for the day, I seem to sit here for so long."

"What are you working on?"

"Every morning, it's the same thing. I have two clients that want me to give them an output based on previous day's results. I have to input their results, run the processes, output the new results and feed them back to the clients so they can make some decisions. Then they let me know what I should do next. I run some more processes and give them the output and some advice. It takes hours of the day and it is so repetitive."

"Yes, repetitive, but it doesn't sound that bad," I said. "Not exactly a rut, well, at least not a deep one. And you must be happier now that you don't have to do a two-hour commute on the bus every day to work in an office. Now you are your own boss. And making more money as well, I would imagine."

"Yes, I am, and I know I have you to thank for that. If you hadn't shaken me up and told me to start my own business, I would still be sitting on the bus, day after day, and working in an office, cheek by jowl with thirty other people who were just as miserable as me, working to make my bosses rich. You know, if they had let me work from home, I would have even considered staying, but they felt the need to breathe down the back of my neck all day long, even though they knew I was working far harder and produced more output than they had ever hoped."

"Well, now you don't have them breathing down your neck, instead you have your pets shedding their fur all over you. A much warmer and cuddlier feeling."

My friend laughed and pushed a disgruntled cat gently from his lap. "I think I'm ready to hear your word to the wise. Or maybe your words to the not quite so wise," he admitted. "I might need a few more words than just one."

"You're looking for transformational learning?" I asked, laughing. "Are you sure?"

He mock-shuddered a little. "Let me make you some tea," he said, "or would you prefer some juice. If you're hungry, I have cake."

"Tea would be nice, since it's a bit cold out today, and I could handle some cake. I've been working out a lot this week and could use a few extra calories." I followed him into the kitchen and watched him.

"You know, thinking about this brings something to mind that might help you with your work problems."

"Have mercy," he said. "Remember I am but weak brained."

I laughed. My friend was actually just short of genius in what he could do once he wanted to do it, but he sometimes needed me to set him off on his way when he became complacent. "Just thinking about inputs and algorithms," I told him.

"What?" he asked.

"Inputs and algorithms. Building models," I repeated. "Thinking along those lines might help you with your work. Instead of taking the time to do the same thing every day, you could create a model that would run in seconds. You can change parameters based on the clients' needs, those are the inputs, and run the model again. Much more efficient than doing the same thing over and over again. So, my word to you, my one word to the wise, would be Modeler."

"Explain a bit more," my friend said.

"Well, think of it like this. A drink. If the day is hot, prepare a cold drink. If cold, prepare hot. Food. If person is overweight, offer carrot sticks. If person is thin, offer cake. The input parameters are temperature and weight. The output is food and drink. Or, think about a smart house."

"My house is just a regular one," my friend joked. "I'm not so sure it is up to an intelligence challenge."

"Very funny," I said. "A smart house anticipates. You can set your coffee machine to start working at a certain time every morning so that it is ready for you. Set the thermostat to turn on the heat just before you get home based on the time of year or the outside temperature. Set your supper to start cooking as soon as you drive up the driveway, even have something started for guests when people drive into the driveway. The point is, it is all about automation, timing and choices. Things happen at different times and depending on different things. You set it all up once and then it runs in perpetuity."

"As I said, my house may not be up to the challenge and I may not be either. I don't know if I want the complication. Are there lots of manuals?"

"No, but that's not the point," I laughed. "I'm not suggesting you get a smart house. I'm suggesting you use the tactic of input and algorithms for your work. Your clients give you certain kinds of input, right."

"Yes, always the same kinds of things. Just different values in the tables of numbers every day."

"Right, and then based on what the numbers are, they want different kinds of things done to the values, correct?"

"Yes. If the numbers are above a certain level, or the data is spread differently, they want different kinds of solutions applied."

"Exactly. Now, if you set up a model, with the input parameters and then ran algorithms, I mean, routines, to do different things, based on the inputs, you could run the model in seconds and run it in all kinds of ways depending on the inputs and the interim outputs. It would take you no time at all. Then you could spend your time..."

"I could work on new ideas that I have."

"You could. Or you could use the work flow to show your clients how they could do the work for themselves and then move them onto more adventurous routes for their business. Or, I was going to suggest you could pet your cat instead of just being its cushion, or walk your dog."

"Very funny. Now, show me how. I think I can manage this kind of change. Will it be transformational?"

"That depends on how I decide to tell you about Graphical Modeler," I laughed. "First the cake and the tea. Then I'll show you. And I think you're right. Transformational will be the key. Let's make sure you don't get into another rut. I don't mind coming to your rescue, but I think you're at the stage where you can rescue yourself."

You will learn how to use the Graphical Modeler.

1. While you are learning about Graphical Modeler, think about transformational learning and make some notes to explain your thoughts.

- How would you advise the person who is having difficulties so the learning curve can be maintained through self-motivation?

- How do you learn?

- How do you stay motivated?

- What stops you from learning new things?

- What takes away your motivation?

2. Read about the Graphical Modeler in the *User Guide, QGIS Processing Framework: The Graphical Modeler* and in the *Training Manual, The QGIS Processing Guide: Starting with the Graphical Modeler*.

3. Follow along with the example of creating a model, below, to see if you have understood how to work with the Graphical Modeler and to get some practice.

4. Create a new model. To do this, you will pick three algorithms from the Graphical Modeler interface. You'll need to create some data or select some data you have already downloaded that suit the algorithms. As you make your choices for your model, consider transformational learning. If you make a lot mistakes, you learn more. If you struggle and have to search for information, read, think, try again, your success will be memorable. Simply finishing the exercise to get it done, doesn't assist you with learning or learning how to learn. See if you can learn more about how you can learn better while you create your model.

Example

Create two shapefiles, one is point, one is polygon. Fields are created for both as shown in Figure 9.1.

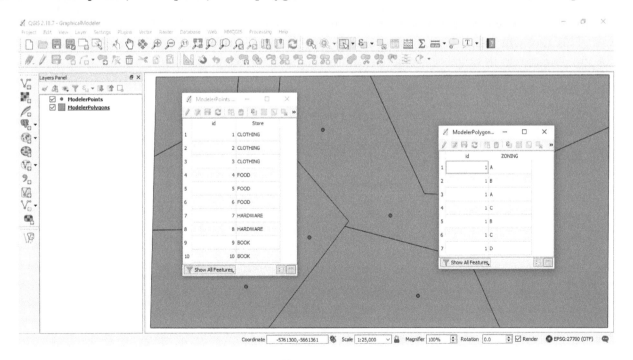

Figure 9.1:

Open *Graphical Modeler* from the *Processing* Menu item.

Add the two shapefiles from the *Inputs Tab/Vector Layer*. Drag and drop them onto the canvas. Fill in the parameter definitions (Figure 9.2).

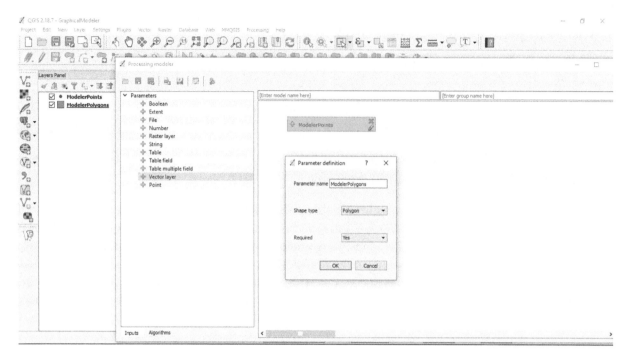

Figure 9.2:

Switch to the *Algorithms* tab, at the bottom of the left panel.

In the *Vector geometry tools* list of algorithms, click on *Fixed distance buffer*. Drag and drop it onto the canvas. This tool will behave in the same way as the tool under the Vector Menu Item, Geoprocessing. Fill in the parameters to buffer the points (Figure 9.3, on the next page).

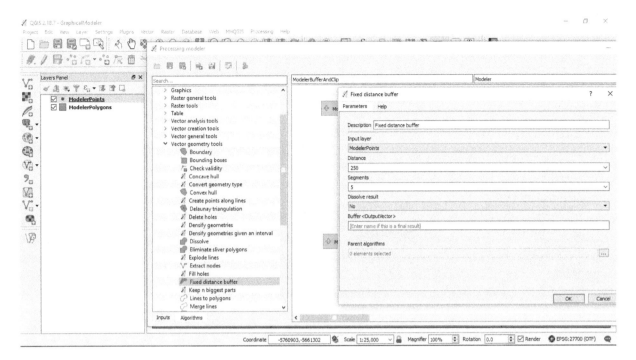

Figure 9.3:

Drag and drop *Clip* from the *Vector overlay tools* onto the canvas. Use the buffers to clip the polygon layer (Figure 9.4).

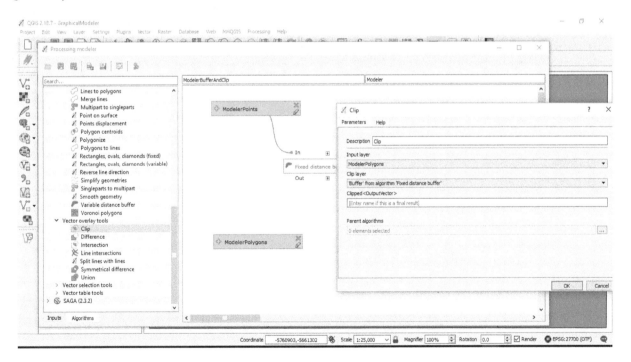

Figure 9.4:

Now the polygon layer has been clipped with the buffers. The final step is to select only the clipped sections that fall within Zoning A. From the *Vector selection tools* list, drag and drop the *Extract by attribute* Algorithm onto the canvas and fill in the parameters (Figure 9.5, on the next page).

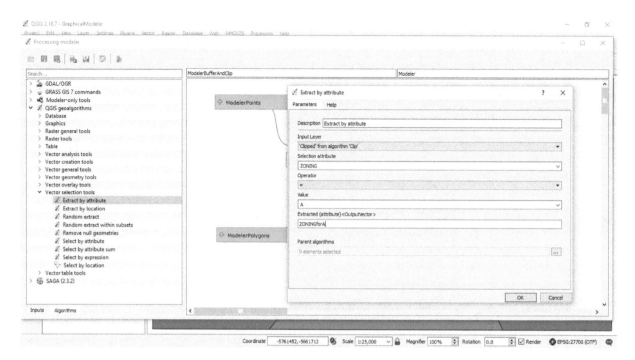

Figure 9.5:

Save the graphical model. Note that you can save it as both an image and a script. Run the completed model (Figure 9.6).

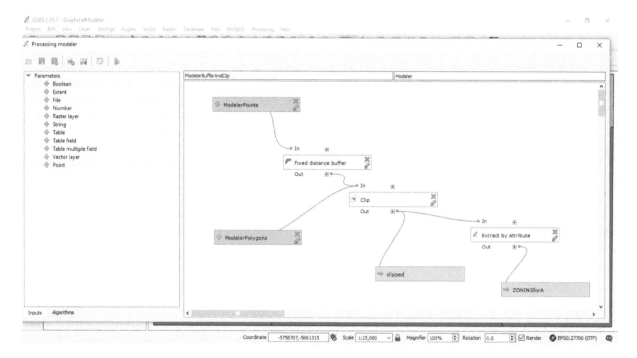

Figure 9.6:

Close the model and check the results. Note that only the polygons Zoned as A have been extracted to a new shapefile (Figure 9.7, on the next page).

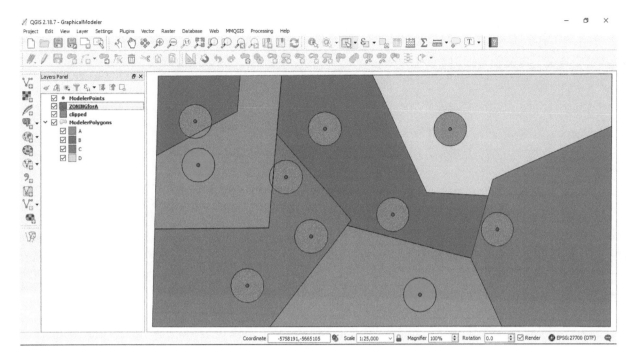

Figure 9.7:

You may want to build the model piece by piece. Run a section, check that it works and add more. You can save the model, close it and open your saved model from the *Folder* icon on the Graphical Modeler. You may want to build the whole model and, if it does not give you the results you expect, you may want to deactivate portions of it so that you can run it step by step, by right clicking on an algorithm in the canvas and selecting *Deactivate*.

The more experienced you are at using the tools in QGIS, the faster you will be at building models, but models will save time once they are built, even if they take a bit of time and effort to build.

Your task is to create a model using at least three different algorithms. Export your model as a graphic and save your QGIS project so that you can show your work.

What will you choose? Have fun with your choices. Everything you do can teach you something. Be bold and adventurous. Now is the time to learn as much as you can!

10. The Three Challenges

Learning Tip Ten: Research shows that having internal motivation is more effective than external motivation. If you really want something, can you rely on your internal motivation? Cultivating its strength will help you in everything you do, not just in learning.

I was doodling on the margin of my notebook, periodically checking the clock, waiting for lunch. I finished the assigned work about ten minutes into the class and had to wait another forty for the bell. I was fast. My work had always received fairly good grades. Good enough to make my parents happy and to be in the top part of the class, keeping my teachers happy. What more did I need to do?

The doodle I was working on had spread over the margin where it started and was threatening to eliminate any note taking space on the page. It was a good one, I thought, symmetrical without being too orderly, interesting enough to focus attention on the detail. Doodling filled the time. Maybe I could become a doodle artist and make millions, I thought. Just being good at school might not equate to making millions and being able to afford everything that I would like to own. I knew I would have to negotiate the transfer of skills between succeeding at school and succeeding in the world after school.

Okay, maybe doodle art wouldn't quite meet all my expenses, but there was lots of time to think about how exactly I would afford the lifestyle I yearned for. My parents were always excited about my future, encouraging me to be thinking about careers, and my teachers were always on about it in class. Way too much thought about the future, I thought. Time for that later. Right now, doodling and waiting for the bell occupied my mind.

"Just before we go, class," said my teacher, interrupting my reverie, "I want to remind you about our conversation at the beginning of this term."

I thought, what conversation was that? There were announcements every day to encourage our learning. Motivational speeches abounded. I had to admit my teacher was actually quite talented at it, trying to interest us from every angle in how to learn and why. I had a grudging respect in that I almost, sometimes, nearly felt motivated. The teacher was gifted in persuasive strategies and I was interested in analysing how it was done. I kept doodling as the teacher continued, but I was listening.

"At the beginning of the semester, I mentioned that just doing your work and handing in your assignments would be enough to guarantee a good grade for this class. If you did your work you could pass the exam without a problem, I am quite sure, but I did mention that doing extra work, not just to complete the assignments, but going beyond, challenging yourself, might bring extra rewards. Well, now I would like to share more about that with you.

"There has a been a very large donation from an industry leader who believes in rewarding students who have gone beyond just fulfilling the requirements. This person thinks that learning for the sake of learning should be rewarded and, accordingly, has set up a contest, for all of the school districts in our area with some amazing prizes."

I stopped doodling. Maybe this was interesting enough for my full attention. Noting the audience reaction, the teacher was smiling, no doubt pleased to have everyone listening for a change.

"I have a printout here of the rules and prizes for what is being called *The Three Challenges*. Pick one up on your way out the door if you think you might be interested. After you have read it, if you think you are a likely candidate, come and see me. As you will read, it will require you to make some choices about how you proceed with completing this class."

The bell rang. I packed up my decorated notebook and walked past the teacher's desk on my way out the door. I wrinkled up my forehead, wondering what the teacher would say to me if I picked up a form, wondering what the teacher would say if I didn't. I didn't have to go beyond wrinkling because the teacher handed me the papers. "Have a look. Maybe you have a chance. Come talk to me about it, if you think you want to give it a try."

"Thanks," I mumbled and took the papers. Behind me, I heard lots of chatter and the rustle of paper as my classmates took papers as well.

Time for lunch, I thought, listening to my stomach rumbling, but instead of going to the canteen, I went around to the back of the school where there were some benches. Not too many people sat on the benches even though it was nice to be outside during the day, just to breathe air that hadn't been filtered through several hundred other pairs of lungs, if not to enjoy the sunshine and the flowers that the school had planted in an effort to make the school look less like a detention centre. I arranged my lunch containers beside me and ate on autopilot while my attention was on reading the papers I had just received. I read:

The Three Challenges – are you up for it?
A story about my observations on learning

If you are reading this, you are likely interested in competing for a prize. That's external motivation. You work hard, you get a good grade, a pat on the back, a prize. You get used to that kind of work and reward system in school. Later, you get a job, and if you work to the rules, you get a promotion with perks and increased salary. External motivation is what drives most of us to try hard. But what would we do if there was no reward for hard work? What would we learn if we didn't have to pass a test? Would we still work as hard as we could? Would we still try our best?

People who are internally motivated want to do well just for sake of doing well. They don't need or want a reward. People who want to learn for the sake of learning are lifelong, continuous learners. Their learning is for personal development. Their hard work is because they care about who they are and what they do.

I believe in personal development and in internal motivation. As such, I also want to reward people for having those characteristics. My goal is to break the mould for the usual education pattern of being told how to do something, practicing it, memorising it and regurgitating it to pass a test. I realise it is somewhat contradictory to give a prize, which is an external motivation, for internal motivation, but I hope you, as a potential contestant, will forgive me and that the reward I am offering to three lucky winners will encourage you to develop beyond what is required for success in your school year, that it will break the pattern and help you to develop skills beyond what is required to pass your class this semester. To take the challenge, you will need to be internally motivated and creative.

Now, what are the three challenges all about? Let's start at the end and go back to the beginning. Let's talk about the prizes. Experience is a great teacher and since I am interested in learning, I am interested in providing experiences that will stretch the winners to their limits and beyond. Accordingly, I have provided funding for three winners to learn new skills over the summer break. All accommodation, food, travel and expenses are covered. One prize will be three expeditions to locations around the world, developed for students of your age, led by a famous geographic foundation. Another prize will be a summer of mentoring at three businesses, headquartered in three different countries, specialising in the development of three different kinds of software. The third prize is a wild card prize, which will be revealed to the winner who receives first place. Are you interested? Do you think you have what it takes? Read on.

All of you have been studying this semester using QGIS. No doubt you have found at least some of the exercises to be interesting or challenging, but did you go beyond what you *had* to do? Did you challenge yourself to do your very best? If you did, great. You're the kind of person I want to hire to work at my company when you graduate. I applaud you and want you to receive one of the prizes by showing me what you can do. If you didn't challenge yourself during the semester, could you change your ways and do it now? If you can, show me.

I have no doubt that most of you who are reading this are capable of winning. I believe everyone has an equal chance, that we all have a considerable amount of talent if we choose to use it. Why shouldn't you be a winner if you try hard? But you have to convince the judges that you are a winner. This is what *The Three Challenges* challenges you to do.

You will be given three tasks in QGIS. Your teacher has gone through a training course to enable the selection of likely candidates from their class after the first round. To go on to the second round of selected students, you will spend three days at my corporate headquarters, discussing and defending your work. From this round, a third selection process will involve you answering questions from your peers, teachers and industry representatives. Still interested, read on to find out what you actually have to do.

First, you will investigate three QGIS plugins, load them and use them to their best effect. You will write about why you selected them, how they work and what they can be used for. You will produce a map illustrating the use. Impress me with your choices and your application.

Second, you will use tools on three different toolbars that you have not used during the semester, based on the text that has been provided for the class. Illustrate how the tools are used with screen captures and tell me why you chose the tools. Amaze me with your skills.

Third, you will use three different menu items that challenge your abilities. Some menu items are easy to use, some are more difficult. I want you to stretch yourself and tell me how much you have learned, so be sure to pick ones that require your investigative skills. Write about your selected menu items and how they work, in your own words. Illustrate their use with mapped data.

Your teacher will have told you that choosing to enter the contest will affect how you complete this semester for your class. As I wrote above, I am interested in challenging the usual education model of examinations. Instead of writing an exam for this class, participation in *The Three Challenges* will be in lieu of a final exam. It isn't easy, but it does require motivation and creativity, two skills that will see you in good stead in the future. Good luck. I hope to meet you soon.

When I finished reading, I saw that my lunch was gone. I wondered what it was. I had been so interested in *The Three Challenges* that I hadn't even noticed. That told me something. When reading was more interesting than food, there was some merit to the content. I stared off into space for a few minutes, thinking about the contest. Thinking about winning. Thinking about the motivation of the person who sponsored the contest. I could see the point. After all, I was certainly tired of memorising facts or processes that I could look up on the Internet. The exams I had to write were mired in the Middle Ages when memory was all people could rely on. With facts readily available to look up, why did we need to memorise them? Were the exams really preparing us for life after school? Wasn't it more important to learn how to think? I agreed with the sponsor, being creative and being motivated to learn new things would likely be more important to my future. Maybe even more important than creating the perfect doodle. I decided to give it a try.

Your task is to complete the challenge.

1. Select your plugins, tools and menu items. Produce evidence that you understand them with screen captures of your application of them in QGIS. Write about how you learned, what you learned and why it matters.

2. If you won second or third prize, where would you go, where would you be mentored. If you won first prize, what would your prize be?

3. What do you think about the incentive of a prize as a method for learning?

4. Is having learned a reward in itself?

5. Bonus question: Outline your own idea for a plugin. What should QGIS be able to do that it can't already do quickly, or at all?

Glossary of Terms

Terms of special significance are indicated in *red*.

Active Layer:

Only one layer is active at a time. When the layer is clicked on, it is highlighted and becomes active so that you can work with it. See Layer.

Active Learning:

Having someone tie your shoes for you is easy. If you remember learning how to tie them you probably remember being frustrated. It wasn't easy! Similarly, having someone tell you the answer or tell you what and how to think, is easy while reading, exploring, investigating, researching and being responsible for your own learning is hard. But will there always be someone there to tie your shoes for you? No. Do you want there always to be someone there to tell you how and what to think? Learning how to learn is an important part of active learning and it takes practice to get good at it. Sometimes the more difficult the task, the more challenging it is, the more fun it can be to accomplish it. Active learning can be frustrating, it can be challenging, but it can be fun as well. It is student centred, it is all about you!

Active Tool:

Only one tool is active at a time. The Active Tool is highlighted. On the map, the cursor becomes an image of the icon that is active. To select another tool, click on the tool and it will become active, while the formerly selected tool deactivates. It is a good idea to check which tool is active before clicking on the map. See Tools.

Algorithm:

A set of steps, or instructions, that is run as one routine or function.

Aggregated data:

Data that is collected from individuals is expressed in groups. Instead of knowing that teenagers live in a particular house, aggregated data will report that a certain number of teenagers live in a particular city. This protects privacy and makes the data more meaningful for analytic purposes.

Annotation:

The text on maps. Text annotation can be added using the icon in the QGIS GUI or using the Label in QGIS Print Composer.

Attribute Table:

The maps are what we usually look at in a GIS, but the tables are what cause the maps to be drawn. Data can be created in a GIS by digitizing onto the map. What is drawn is converted into data in an attribute table which stores the location of points, or the points along a line or of the boundaries of a polygon. For example, an attribute table of a point shapefile for Cities, will contain the location of the city as an X,Y coordinate and could also hold attributes of the elements which describe them: a field for Name, Population, and Size, if you add the fields. See Field.

Bookmark:

A saved location viewed at a specific scale that can be retrieved.

Buffer:

> A Buffer is an area around a feature that is calculated based on the distance specified from the feature. Buffers are polygons. They can be drawn around points, lines or polygons.

Centroid:

> The middle of a polygon.

Classifying Data:

> Maps are a form of communication. We want our maps to communicate information clearly and in an easy way for the user to interpret. If we simply put all kinds of numbers on a map, it wouldn't be easy for a reader to use, but if we group the numbers into a few classes, for example, 0–10, 10–20, and 20 and above, and use colours or size to illustrate the different classes, it is much easier to see. Sometimes we know what we want to show, sometimes we need to explore methods to uncover what the data holds. Sometimes the data is limited and can only be displayed using certain methods. There are many different ways to group, or classify information and they depend on the data and the intent of the person creating the map.

Clip: See Overlay Analysis

Concept Map:

> A Concept Map is a way to display information using a graphic representation. Concept Maps can organise information, structure thought, indicate potential new relationships, and integrate old and new information. Looking online at images of Concept Maps is a good way to understand how they work and what can be done, but your Concept Map is uniquely your own and is an illustration of what you think is important about a particular concept.

Coordinate Reference System:

> Coordinate Reference System are geographic or projected. Geographic systems represent the world as a globe and use latitude and longitude in degrees, minutes and seconds, or decimal degrees. Projected systems represent the world in two dimensions, like a paper map, which stretches and distorts shape, area, distance, or direction. The use of the map dictates the choice of the projection so as to preserve the most important characteristics. A popular choice of projection is the Universal Transverse Mercator in which the X coordinate is represented with six numbers (called an Easting) and the Y coordinate by seven (known as Northing). See X, Y Coordinate.

CSV:

> Comma Separated Value. A text file, rather than a spreadsheet, which can be read as a spreadsheet by software. The commas represent the breaks between data in different columns. The first row of data is usually the column headers.

Difference: See Overlay Analysis.

Digital Elevation Model:

> A DEM is made of cells and is a Raster. It is important to note that this is a model of the elevation. Models are approximations of reality. Each cell of data has a height (usually called a Z value) associated with it, as is captured by a satellite or airplane flying over the location. Sometimes used synonymously with DTM. See Digital Terrain Model and Raster.

Digital Terrain Model:

A DTM is made of points and/or lines and is a vector elevation model. It is important to note that this is a model of the elevation. Models are approximations of reality. Each point of data has a height (usually called a Z value) associated with it, as is captured by a satellite or airplane flying over the location. Sometimes used synonymously with DEM. See Digital Elevation Model.

Digitize:

To digitally draw or trace.

Dissolve:

See Overlay Analysis.

Distortion:

In digital images, distortion manifests as pixels not being in the correct relative location to other pixels in the image and not being in their correct geographic location of a projected geographic coordinate system. This can make an image look strange, like looking at your face in a spoon. Distortion occurs during the collection process. It can be removed using a variety of techniques so that the entire image lines up with what is on the ground.

In projected maps, there is always an element of distortion. Cartographers will need to prioritise what element(s) need to be preserved, from among Shape, Area, Distance and Direction, as not all can be maintained. (See Orthorectification)

Feature:

One element of a Layer, either vector or raster. Features can be added, moved, identified, selected, and deleted using tools. See Layer, Vector, Raster, and Active Tool.

Field:

A field is a part of a table, also called an Attribute Table. Depending on the program, fields may also be called columns. Each field has only one type of data in it which can be text (also known as string), numerical (integer or real), or date. A text field can accept numbers, but a numeric field cannot accept text. Be sure to set the field type correctly for the attributes you need to store. See Attribute Table.

Field Calculator:

Button in the Attribute Table that allows changes to existing fields or the creation of new fields based on functions similar to those in a spreadsheet. Works on both text and numeric values.

Georeference:

Paper maps can be scanned and georeferenced, that is, once they are digital, they can be displayed in QGIS in the correct geographic location. To georeference a map, data from the map, such as grid reference marks and numbers, can be used. Select a point on the digital image and then click on the place where it should be in the QGIS map display. At least four points should be collected.

Geocode:

A process whereby geographic coordinates are assigned to text such as addresses.

Geoprocessing:

A set of tools that work on GIS data. Geoprocessing allows for GIS to create new, derived data from existing data sets. See Overlay analysis.

GIS:

Geographic Information System. A GIS is a system that works with spatial data, to store, manipulate, display, and analyse. It's more than just a digital map. It's a database with topology, meaning that each data element knows where it is in relation to other elements. As with any database, the data can be queried, and, since GIS has spatial characteristics, it can be queried about what it is and where it is. GIS started in the 1960's, but did not become mainstream until recently. Now we use online GIS web maps all the time to plan our trips and to find locations without really knowing much about the GIS that is working in the background. Knowing more about GIS may help us to make more fully informed decisions about our world.

Graphical Modeler:

Tools can be run one at a time, or they can be run as a sequence in a group. Graphical Modeler allows for setting up a sequence of tools to run with one click, like a chain reaction. The output of one part of the sequence provides the input for the next. Changes to input parameters can be made easily so that different scenarios can be tested quickly.

Group Work:

Working together as a group in school paves the way for working as a team in jobs. From the amount of literature available and the number of companies that offer services to businesses on how to develop effective teams, it would appear that we don't always learn how to work as a group very effectively while we are in school and we still need help with it when we start to work. Why is that? Do you think groups or teams most often fail because they don't know what they are supposed to do, because they don't know how to do it, or because they can't/don't communicate with each other? Or is it a combination of all three? Do you think learning to work as a team before starting the work the team needs to do is a good idea? If you are interested in more about team work, search online for Belbin's Team Roles and find out what kind of team player you are.

GUI:

Graphic User Interface. The icons, menus, and elements that display on the computer screen and respond to being clicked by the mouse.

Heat Map:

A map of an area that illustrates intensity with colour.

Help Files:

If you don't love help files, if you don't read them regularly, read this! You might think that Help Files are too difficult to read, or that it is too difficult to find the information you are looking for when you use them, or that they are just boring. For some Help Files, this is likely true, but there are so many to choose from that you can usually find one that meets your needs. When you search on the internet, you are really using a gigantic Help File and you likely find what you are looking for after you have a look through a few sites and figure out how to phrase your question so that you get the file you need. You get better at using Search Engines as you practice. You'll get better at understanding different kinds of Help Files in the same way. Searching for, reading and even writing Help Files (after all, aren't your notes just another form of Help File) is a skill, maybe even an art. It gets easier as you get more experience. Evaluate Help Files. What makes one better than the other? Write your own Help Files. How should they be written?

Hypothesis:

A guess that will be investigated to see if it might be correct.

Interpolation:

The estimated value between known values.

Intersection:

See Overlay Analysis.

Join:

Not all data that is usable in a GIS has a location in the world. A spreadsheet of data that does not have a geographic field can be joined to geographic data if the two datasets have a common field.

Layer:

GIS organises information by type. If the element of the world can be modeled as a point, such as the location of a tree, then all of the points that represent trees can be put into one file. When this file is displayed in the GIS, it is called a layer. Other layers will represent other files of different kinds, such as polygons for lakes or lines for roads. Layers have a display order, with the topmost layer drawing onto the display last, so that it draws on the top of all of the other layers. The layer is only a pointer to the data. It is not the data itself. If you move the location of the data on your computer, the layer will no longer draw. See Active Layer.

Layout:

A layout consists of a map or maps and marginalia (information about the map) such as the scale, legend, title, author, date, and projection. The Print Composer holds the Layout.

LIDAR:

RADAR uses sound waves to map surfaces. LIDAR uses light pulses to do the same. The high density of the points used to map areas means that a highly detailed map can be produced. Previously unknown geographic and historic features have been discovered through the use of LIDAR.

Map Algebra:

Pixels in raster data have numeric attributes associated with them. Overlaying multiple raster layers, the numbers in each pixel can be used in algebraic operations to produce an output raster dataset.

Metadata:

Data about data includes such elements as date produced, how it was produced, author of the data and restrictions on the use of the data. Using data without a metadata file attached to it is not advisable.

Mercator:

Map projections allow for the spherical world to be transferred to flat paper. There are three main ways that a projection can be made. The paper can be made into a cone and placed over the sphere, the paper can be laid flat against the sphere or the paper can be wrapped around the sphere resulting in a cylindrical projection. Mercator is a cylindrical projection first developed by a cartographer seeking to assist navigation with a map that would not distort direction. Area is distorted, so Mercator projections are not suitable for maps of the World, although they were frequently used for this purpose.

Model:

A model is a representation of reality. Models are often used in GIS to investigate What-If scenarios. It saves time to build a model, change some parameters and run it multiple ways, to predict what might occur in different circumstances. Care must be taken, however, to remember that a model is only a representation and a prediction, it is an educated guess, but not a guarantee. See Graphical Modeler.

Nadir:

The point directly below a given location. For example, if a satellite is taking a digital image of an area spanning 25 km by 25 km, the nadir will be at the exact centre of the area.

Node:

A point on a line, or on the perimeter of a polygon, where the line changes direction. Also known as a vertex.

Orthorectification:

A process which removes distortion from a remotely sensed image, such as an aerial photograph or a satellite image, so that all parts of the photo are at the same scale. (See Distortion)

Overlay Analysis:

A GIS uses layers of information. Each layer is of a certain type and stores attribute information particular to its characteristics. Overlay analysis allows for the combining of layers to produce new information, both descriptive and geographic, and includes the following tools:

> *Clip*: A geoprocessing tool which requires an input shapefile (to be clipped) and a clipping shapefile (acting as an outline for the clip required). The action is similar to a cookie cutter on dough in which you want the cookie and not the rest of the dough. The process will produce a new shapefile. The clipping shapefile needs to be of type polygon, but the input shapefile can be of any type.

> *Difference*: A geoprocessing tool which requires two polygon shapefiles to be input. The second shapefile erases the area of the first where it overlaps. This differs from Clip which gives a cookie. Instead of acting as a cookie cutter, you make a doughnut. The process will produce a new shapefile.

> *Dissolve*: A geoprocessing tool which works on one shapefile. Polygons in that shapefile that are adjoining, and have the same data in the attribute field, will be merged. Or, all polygon boundaries within the shapefile could be merged into one. The process will produce a new shapefile.

> *Intersection*: A geoprocessing tool which requires two shapefiles to be input. The process will produce a new shapefile. The output shapefile will contain only the features in common to both input layers.

> *Union*: A geoprocessing tool which requires two polygon shapefiles to be input. The process will produce a new shapefile. The output shapefile will contain all the features in both input layers.

Pan:

To move to a different area of the map while staying at the same scale. See Zoom in, Zoom out, and Scale.

Plugin:

Plugins provide additional functionality. They are accessed via the Plugin Menu, Manage and Install Plugins. Plugins used in this book are: *Open Layers Plugin, Geosearch, Zoom to Coordinates, Azimuth and Distance,* and *Qgis23js.*

Project:

A Project is a workspace that is created and can be saved as a .qgs file. It holds Layers and the properties that have been saved for them as well as Print Composer files of map layouts. Anything you do within a Project can be saved and will be retrieved when the Project is reopened.

Projection:

See X, Y Coordinate and Coordinate Reference System, and Mercator.

Property:

Layers have properties. Double clicking on the Layer brings up the layer properties. Right clicking on a Layer brings up a menu list, one of which is properties. The types of properties for a Layer are dependent on the kind of Layer it is. Different Layer properties allow for alternative ways to display data including such things as labeling and symbology (colour, size etc.), and will detail the location of the file on your computer that is being referenced by the Layer file. See Layer and Active Layer.

Pseudo colour:

Pseudo means false, so pseudo colour is false colour. In a raster image, each cell has a different number, from 0 to 255, which maps to a colour for a colour image, or to gradations of grey for a black and white image. To make the gradations of grey more apparent in a black and white image, pseudo colour can be used.

QGIS:

An Open Source GIS that can be used free of charge. Open Source software is developed and supported by a worldwide community of users who are dedicated to providing a complete, user friendly alternative to commercial software. In 2011, it is estimated there were 100,000 users of QGIS worldwide.

Qualitative Research:

Qualitative data is not quantifiable. It relies on descriptions, not numbers. Sample sizes for the research are small. Researchers do not know what they will find when they begin; they seek to uncover information. The studies are not generalizable; they cannot be used to describe larger populations. The researchers are interpreters and observers.

Quantitative Research:

Quantitative data involves numbers. This is the kind of research most commonly associated with science. Researchers start with a hypothesis. The studies use statistical methods and rely on large, random samples. It is hoped that findings can be used to describe larger populations or events.

Raster:

Raster data is a way of modeling the world using a grid of equal sized cells in rows and columns. It is considered a continuous method of data representation and is the alternative to Vector data. Every part of the map is filled with data, for continuous coverage, not just parts of it as is the case in Vector's discrete representation. Each cell can only represent one value, so the size of the grid's cells is extremely important. The size of the cell represents the raster's resolution. Whatever fills the majority of the cell is what gets represented. If one cell has a house in 75% of the cell and trees in 25%, then you won't see trees in that cell. If you zoom in closely to a photo from your mobile or digital camera, you will see the cells of the raster, called pixels (short for picture elements). See Vector.

Render:

How something draws or displays.

Resolution:

A numeric value that describes how much of the earth is represented by one pixel of a raster. If the resolution is 30 metres, then anything under 30 metres will not be detected in the raster. Where areas are homogenous, a coarse resolution will suffice. Areas such as cities, where features are small and areas are heterogenous, need a fine resolution of at least a metre, preferably less. A fine resolution shows more detail but is more expensive to produce and purchase.

Scale:

Map scale is a representative fraction. A map of $1/2000$ (or 1:2000) means that 1 unit on the map represents 2000 units on the ground. A map of $1/20,000$ (1:20000) means that 1 unit on the map represents 20,000 units on the ground. Considered as a fraction, one 2000th is larger than one 20,000th which is why a 1:2000 map is a larger scale map and a 1:20,000 map is a smaller scale map.

Scale Dependency:

Entire layers, or parts of layers, such as labels, can be set to display within a certain scale range. This allows for information to become available to the user when they will find it useful without cluttering the map. For example, when you look at a map you might not see all the information, but as you use the zoom function to see a particular area more information will appear as you zoom in. For example, labels for minor streets could show at a scale that is bigger than the labels for major streets. Local streets may not display at a smaller scale where major highways are the focus.

Shapefile:

A format for saving digital data representing points, lines and polygons with a special attribute that stores location. The shapefile will appear as one layer in a GIS, but is made of several files stored on the computer, .dbf, .shx, .prj, and .shp. All files must be present and in the same location for the shapefile to be functioning. Shapefiles can be created (Create New Shapefile Layer) and can be added (Add Vector Layer) to one or to multiple projects. To add a shapefile that has already been created, browse to the file location and select the .shp file. See Layer.

Snapping:

A specified distance can be set so that, for example, the end of one line will snap to another if it falls within that distance. This assists with the creation of accurate shapefiles.

Spectral Reflectance:

When light (the spectrum) hits an object on the earth, some of it bounces back. How we see an object depends on what is reflected back from it.

Supervised classification:

A raster dataset can be changed into a vector dataset to provide, for example, a land use map. The operator identifies a pixel as being, for example, coniferous forest, or an urban area. The software will run a process on the raster so that all pixels with similar values will be given that identity See Unsupervised classification.

Symbology:

The ways that data can be visualised including such elements as size and colour of font; colour, pattern and outline for polygons; or thickness, colour and pattern for lines. Symbology is an important part of cartography which assists with communicating information through maps and is both a science and an art.

Tools:

Tools are icons on the GUI. Tools are organized into toolbars. Hovering the cursor over a tool will reveal the name of the tool. See GUI and Active Tool.

Topology:

GIS allows for topological relationships between objects, such as adjacency or connectivity. Topology allows for features within a GIS to "know" about other features. Topology rules ensure that features will adhere to specifications such as one object contains another, all polygons are closed, all lines are snapped to other lines. Having topological rules assists with keeping data accurate.

Transformational Learning:

Think back. When you suddenly understood how to tie your shoe laces, or understood how the decimal system worked, not just when you could follow the steps to tie your shoes, or that counting to 100 was just counting to ten ten times. Aha!, you said, now I understand. The world changed for you. That was transformational learning. Sometimes you might firmly believe one thing and then something disrupts your thoughts and you see the world in a different way. That is also a moment of transformation. These moments aren't easy to come by and can sometimes cause discomfort in the person who is on the cusp of change, but they are thrilling moments. Transformative learning always brings something exciting into our lives. Look out for opportunities to experience it!

Triangulated Irregular Network:

A TIN is a vector data set and is a kind of Digital Terrain Model. Points with X, Y, and Z values are used to create a TIN. X and Y represent location and Z represents elevation. The points become the vertices of triangles of varying sizes and shapes. The triangles don't overlap. See X, Y Coordinates and Digital Terrain Model.

Union: See Overlay Analysis.

Unsupervised classification:

A raster dataset (see Raster) can be changed into a vector dataset (see Vector) to provide, for example, a landuse map. Pixels with the same, or similar, values can be grouped by the software to provide a map with areas that are likely to represent one kind of land type, such as coniferous forest, or urban areas. See: Supervised classification

Vector:

Vector data is a way of modeling the world using points, lines, and polygons. It is considered a discrete method of data representation. Depending on scale, rivers are lines or polygons and cities are points or polygons. Vector data is good for boundaries and locations of non-natural elements. Coastlines and swamps can be mapped with vectors, but exact representation of water levels will always be problematic. Imagine standing at the edge of a wetland, how wet do your feet have to be before it has ceased being land and started to be swamp? See Raster.

X, Y coordinate:

On a coordinate grid, X represents the horizontal and Y represents the vertical. For latitude and longitude, X represents the longitude and Y the latitude. As the cursor moves over the map, the coordinates are shown in the Status Bar at the bottom of the map. Sometimes X, Y coordinates also have a Z value associated with them representing elevation. See Coordinate Reference System and Digital Terrain Model.

Zoom in:

To change the view of the map display to a larger scale map while staying in the same area. See Zoom in, Pan, and Scale.

Zoom out:

To change the view of the map display to a smaller scale map while staying in the same area. See Zoom out, Pan, and Scale.

Index

Books from Locate Press

QGIS Map Design

USE QGIS TO TAKE YOUR CARTOGRAPHIC PRODUCTS TO THE HIGHEST
LEVEL.

With step-by-step instructions for creating the most modern print map designs seen in any instructional materials to-date, this book covers everything from basic styling and labeling to advanced techniques like illuminated contours and dynamic masking.

See how QGIS is rapidly surpassing the cartographic capabilities of any other geoware available today with its data-driven overrides, flexible expression functions, multitudinous color tools, blend modes, and atlasing capabilities. A prior familiarity with basic QGIS capabilities is assumed. All example data and project files are included.

Written by two of the leading experts in the realm of open source mapping, Anita and Gretchen are experienced authors who pour their wealth of knowledge into the book. Get ready to bump up your mapping experience!

The PyQGIS Programmer's Guide

EXTENDING QGIS JUST GOT EASIER!

This book is your fast track to getting started with PyQGIS. After a brief introduction to Python, you'll learn how to understand the QGIS Application Programmer Interface (API), write scripts, and build a plugin. The book is designed to allow you to work through the examples as you go along. At the end of each chapter you'll find a set of exercises you can do to enhance your learning experience.

The PyQGIS Programmer's Guide is compatible with the version 2.0 API released with QGIS 2.x. All code samples and data are freely available from the book's website. Get started learning PyQGIS today!

Geospatial Power Tools

EVERYONE LOVES POWER TOOLS!

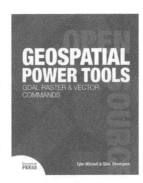

The GDAL and OGR apps are the power tools of the GIS world—best of all, they're free.

The utilities include tools for examining, converting, transforming, building and analysing data. This book is a collection of the GDAL and OGR documentation, but also includes new content designed to help guide you in using the utilities to solve your current data problems.

Inside you'll find a quick reference for looking up the right syntax and example usage quickly. The book is divided into three parts: *Workflows and examples*, *GDAL raster utilities*, and *OGR vector utilities*.

Once you get a taste of the power the GDAL/OGR suite provides, you'll wonder how you ever got along without them.

Discover QGIS

GET MAPPING WITH DISCOVER QGIS!

Get your hands on the award winning GeoAcademy exercises in a convenient workbook format. The GeoAcademy is the first ever GIS curriculum based on a national standard—the U.S. Department of Labor's Geospatial Competency Model—a hierarchical model of the knowledge, skills, and abilities needed to work as a GIS professional in today's marketplace.

The GeoAcademy material in this workbook has been updated for use with QGIS v2.14, Inkscape v0.91, and GRASS GIS v7.0.3. This is the most up-to-date version of the GeoAcademy curriculum. To aid in learning, all exercise data includes solution files.

The workbook is edited by one of the lead GeoAcademy authors, Kurt Menke, a highly experienced FOSS4G educator.

See these books and more at http://locatepress.com

Lightning Source UK Ltd.
Milton Keynes UK
UKHW052346160822
407402UK00002B/4